APPLES

APPLES

Frank Browning

North Point Press

A division of Farrar, Straus and Giroux

New York

North Point Press
A division of Farrar, Straus and Giroux
19 Union Square West, New York 10003

Copyright © 1998 by Frank Browning
Distributed in Canada by Douglas & McIntyre Ltd.
Printed in the United States of America
Designed by Jonathan D. Lippincott
First edition, 1998

Library of Congress Cataloging-in-Publication Data
Browning, Frank.
     Apples / Frank Browning. — 1st ed.
        p.   cm.
     Includes bibliographical references (p.   ).
     ISBN 0-86547-537-7 (alk. paper)
     1. Apples.   2. Apples—Kentucky—Wallingford.   3. Browning, Frank.
     I. Title.
     SB363.B725   1998
     641.3'411—dc21   98-27252
Grateful acknowledgment is made for permission to reprint "Burrow Hill Cider Farm" by James Crowden. First published in *In Time of Flood*. Copyright © 1996 by James Crowden. Reprinted by permission of the author.

Title page photograph, *Apple Orchard at Peak of Bloom*, copyright © 1998 by John Marshall. Reprinted by permission of the artist.

Endpaper art: *McIntosh* and *Cox's Orange Pippin* copyright © 1993 by Elisabeth Dowle. First published in *The Book of Apples*, Ebury Press copyright © 1993 by Joan Morgan and Alison Richards. Reproduced by permission of the artist, c/o Margaret Hanbury, 27 Walcot Square, London SE11 4UB.

The last and frequently most fulfilling pleasure in completing a book is the recollection of those people who have given the author guidance and support along the way. First are those who, believing that such a quirky piece of personal and agricultural storytelling would make a book, have supplied me with a steady stream of clippings, references, translations, and general inspiration, chief among them Sharon Silva, Shirley Browning, Cecelia Brunazzi, Laurie Garrett, Eugene Kahn, Catharine Roehrig, Frank Viviano, and Jeff Weinstein. Several growers and horticultural specialists have been especially helpful, including Ed Fackler, Phil Forsline, Joe Garrett, Yves Lespinasse, Judith and Terry Maloney, Doug Newell, Dan and Donna Rouster, and John Strang. Herb Aldwinckle and Jim Cummins, of the Cornell University/New York State Agricultural Experiment Station generously agreed to read, critique, and correct portions of the manuscript; all remaining errors and oversights, of course, are mine. Librarians at the Brooklyn Botanic Garden, the University of California at Berkeley, the Folklore Collection at the University of London, the Cider Museum at Hereford, England, and the collection at Colonial Williamsburg provided invaluable help. Elisabeth Kallick Dyssegaard has kept the project on track, and Ingrid Sterner saved me from too many overly ambulatory sentences.

For Mary Hester, who has taught me persistence,

and

For Joan Daves, who never stopped believing

Contents

APPLES

Prologue

I was raised surrounded by apples.

My earliest visual memory is of the swaying limbs atop a row of Golden Delicious trees. It was summertime and the Kentucky sky hung hot behind those high twigs, their leaves leathery-green on top, soft as down underneath. Rows of Goldens stood next to a dirt road that separated the orchard from our front yard. Beyond the top row an acre of apple trees marched down the slope toward the gravel highway, the ground beneath them blanketed by rye and fescue and orchard grass, Queen Anne's lace, orange milkweed, and buttercups. The apples, little bigger than shooter marbles, played hide-and-seek with a child's eyes unless, on a June afternoon, a late thundershower drenched them and the last golden sunlight showed them off like Christmas balls. They were in that way both ordinary and magical, as common as toast, as elusive as dreams.

Not until adolescence, when I began to accumulate the images of English and American poets (Yeats's "silver apples of the moon . . . golden apples of the sun" and Millay in her moist earth listening for "drenched and dripping apple trees") did I learn how widespread were the lures and mysteries of the apple. Apples were simply the family trade. A bad apple crop meant the car wouldn't get non-essential repairs and my mother wouldn't make her annual trip to New York to visit her sister. Pedestrian stuff. At the same time, our crop set us apart from the tobacco growers and dairymen who made up most of the county's farmers. We and the other apple grower one ridge away were somehow a touch strange and ever so slightly suspect.

It seemed to have to do with pleasure. No one else in our part of the world grew anything for money that gave so much plain delight. The annual act of taking the first apple, roundish, blushed with crimson, the flesh beneath its skin firm and sturdy and ready to explode upon the tongue, announced by a fragrance so clear and subtle that the memory centers deep within the brain let loose an avalanche of nearly forgotten summer tastes, memories that are themselves inseparable from the expectations of eager youth—this plain act of snapping the fruit from the branch or lifting it from a wooden crate and drawing it up toward the ready mouth was an affirmation of the senses. To eat into the apple, to press the edge of the teeth past the taut unwilling skin into ready white meat, to feel the spray of tart and honeyed juices rain down against the tongue and wash over the palate, was to know again how exquisite are the treasures of the ordinary earth. Apple season—not the apples of summer best left for sauce pots and jelly jars, but the apples that fall in September and October, when the sun is sliding south—is the time for harvest mischief. It is the time of last temptation before the cold and darkness, and the growers are the agents of temptation.

You understood about apples and temptation when the preachers arrived Sunday afternoons.

Along with hundreds of other country and city people, they would rattle up the rutted mountain road to our orchard. Like everyone else, they and their families would park under oak trees, get out of their cars, and amble into our dirt-floor sales shed. No matter how full the outside air was with dust and goldenrod pollen, inside the shed the perfume of fresh apples was overwhelming. "Whew, shore does smell like apples, don't it!" they'd say, and add, "Lands'a mercy. Never saw so many apples!" even though they had been walking into the same sales shed, buying the same varieties for twenty years. Wood-slatted bushel crates of apples were stacked four high and three deep: Jonathans, Cortlands, Grimes Goldens, Golden and Red Delicious, Winesaps, Rome Beauties, and even rare old apples called Winter Bananas, Black Twigs, and King Davids. Each variety was marked with the apple's name and price.

The preachers would walk in after church to choose their apples. The ordinary apple customers would mill about for fifteen or twenty minutes, sniffing, punching, peeling, slicing the fruit, make their selections, and carry their bounty off to their cars and pickups in gunnysacks or big brown bushel bags tied with twine. The preachers would do the same thing, up to the point of purchase. But because it was Sunday, they would ask us to set their apples aside in the rear packing room. Then they would return Monday to pick them up and pay us.

I suppose the authorities could have closed us down for violating Kentucky's blue laws. No one else, except for the town pharmacist, opened on Sunday. The supermarkets were closed. So were the gas stations. By the time I was old enough to ask questions, my father assured me that apple orchards were special. Without Sunday sales, apple growers couldn't stay in business.

As a practical matter, Sunday was the only day most of the country people were free to come buy apples. Until the 1960s, few women drove cars, and their husbands were busy workdays. Farmwork started at six-thirty or seven o'clock in the morning and ended well after sundown. Inside the household, there were also strict regimens. Monday was wash day. Saturday was town day, and each square (there were only two) was packed with men in denim overalls standing around jawing while their wives bought groceries. Saturday night was bath night. Sunday, after church and chicken dinner, was the time "to drive out to the mountain" to get apples. City people who had come from Lexington or Cincinnati took the day as an autumn outing; they usually bought anywhere from a peck to a bushel and a half of fruit. Country people, who were most of our customers, packed off from two to ten bushels, and often they returned a month later for a like amount. So Sunday was the practical day to buy apples. Still, there was something else: because it was Sunday, and because the apples were both wholesome and primordially forbidden, the whole experience bore just a trace of an illicit adventure. Driving up that scary mountain road with the steep cliffs (maybe 150 feet down), spending money just after the sermon was over, letting the kids loose to romp in the orchard, looking over the city folk in their fancy hats and polished Pontiacs (or, for the city folk, pointing at the pickups and listening to the hillbilly twang), and complaining politely about how the prices seemed "awful high this year": all this made coming to get apples unlike any other shopping experience.

Most of our customers grew plums and cherries and blackberries in the back yard or along old fence rows. Bananas and oranges were store-bought food. But apples, at least good apples, occupied a special place, geographically and culturally. To grow them, you had to use poisonous chemicals with odd names—arsenate of lead, Malathion, captan. You also had to have special equipment to spray the chemicals onto the trees. Tobacco, the crop that most Kentucky farmers grew, was little more than a high-class weed whose cultural practices had been passed along from generation to generation. Serious apple growing took book knowledge, which itself brought to mind the lessons of Eden, which of course is all about the invitation to temptation.

Of all the fields that farmers tend, the apple orchard seems to me the most inviting. Others have their own unmistakable beauty. Cornfields betray geometric patterns that can be as mesmerizing as Navaho rugs. Bean fields run on in endless dark-green lines against the sunburnt soil. Wheat fields press their swaying fingers to the wind. Orange groves fluff themselves up like fuzzy columns of deep-green porcupines, but they never show their structure, never invite tomboys to shinny up their trunks and crawl along their bare scaffold limbs. Apple orchards wander along hillsides. They meander around boulders and roadways. They make room for wet spots where trees won't grow. They encourage guests to stop in and visit a while.

Once lured inside, visitors come upon fantasy hideaways full of imaginary furniture. Sturdy lower limbs become fast horses leaping from the trunk. High in the top, twisted branches make crow's nests and pirate perches. Halfway up hang the swaying wooden hammocks of August daydreams, where faces made of leaves and clouds tell tales of impossible voyages. Apple trees promise temptation to the young and fertile mind. Autumn journeys into the orchard seem to stir the memory of all those temptations to lassitude. With winter only a month or two away, the trek to the orchard carries just the slightest feeling of a fling. The preachers knew that, too, and maybe theirs was the finest fling of all, as they stepped inside the fruit shed and then concocted a transparent scheme to avoid a technical violation of the law of the Sabbath.

A few years ago, when my brother and I took over the farm, we began to press cider. Cider is a misnomer in America. Cider, as the English or the French or the Spanish or even our Colonial ancestors knew, is, properly speaking, a fermented drink, the wassail of English caroling, the beautiful, bubbling champagne-looking stuff that Bretons sip with crispy crepes (about which, much more later). But since real cider all but disappeared in America a hundred years ago, the word has been given to fresh, whole apple juice.

Our entry into cider making taught us even more about

apples and temptation. When we started out, no one in fifty miles had made cider for decades. It wasn't part of the diet. The only apple juice our customers knew was the denatured yellow liquid produced by industrial juice bottlers and served to children in school lunchrooms. Yet when we told them we had cider and offered them a sample cup, a sinner's smile would often spread across their faces. "You better be able to drive," the wives told their husbands. Our orchard is in a dry county where Prohibition has remained in effect since the Depression. But the memory of moonshine is strong.

Our county and the county just east of us were two of the best white-lightning spots in Kentucky all the way through Prohibition. Bars and nightclubs in Covington and Newport, down on the Ohio River across from Cincinnati, were among the nation's biggest moonshine consumers. Even into the 1970s and 1980s. Mostly their patrons wanted sour mash or corn liquor. But a good many had a taste for applejack, which our respectable customers understood was just a few days down the fermentation trail from the fresh, vitamin-packed whole juice we drew from a spigot in our converted milk cooler.

One of our neighbors—he lived only four or five miles away—was Asa Muse. Asa was the last of the county's great moonshiners. Asa's grandfather, F. T. Muse, also known as Uncle Toll, had prospered all through the Great Depression on applejack and corn whiskey. F.T. grew corn, much of which he ground and turned into sour mash, and he tended an eighteen-acre apple orchard. Half his apples went into the big copper still that he kept up on the southern tip of Pea Ridge, above his house. Red Delicious, Asa told me, made the best applejack.

Asa had already retired when he told me about his moon-
shining and about how when he was fourteen his grandfather
sent him up the hill into the woods with a .38 special and told
him to shoot anybody who came around. Asa made a lot of
money, he said, in the applejack and moonshine trade—maybe
a quarter million dollars, a quarter of which he allowed he'd
tucked away (but not in a bank). He'd never gone to jail,
though he was tried three times. "In the forty years I made
whiskey, I've been in jail twelve hours. They tore up three
stills, and they'd come and arrest me. I'd get me a good lawyer.
I beat 'em three times." He laughed, his stubbly jowls jiggling
like an old basset hound's. "They didn't like that very well."

Asa's great-grandfather had hauled his family over the
Appalachian Mountains from Virginia and been awarded a
twenty-thousand-acre land grant early in the nineteenth century. They had fought Indians, cleared ground, and established
a mill to grind corn. The village just down the hollow from
Asa is still called Muses Mills. From the beginning, Asa said,
the family had a knack for making liquor. As profitable as it
was, though, and as much land as they held, it all seemed to
slip away—so that by the time I was growing up, the name
Muse carried the sweet-sour fragrance of wildness and decay.

Our apple orchard was three or four miles from the site of
the old Muse orchard. We never cooked up any applejack, and
I was well into middle age by the time I met Asa. Yet I always
felt an odd connection with the Muses. While we were regarded as respectable, educated, and, in local terms, prosperous
and they were the raucous mountain men (nobody ever spoke
of the Muse women) who got everybody drunk on Saturday
night, both the Brownings and the Muses were tainted by their
association with forbidden fruit.

Though others might violate the Sabbath by spending their money on Sunday, theirs was a minor sin, and an easily forgivable one. But as the people who had planted the trees, harvested the fruit, and then offered it, we were the authors of their temptation. In this deeply religious mountain country, where a substitute schoolteacher once eyed a novel on my desk as though it were a satanic cookbook, we Brownings were seen as never too far from sin. Even the children who brought apples to school in their lunch boxes were somewhat suspicious. One became horrified when in sixth grade I explained that we made all our trees with grafting knives and tape. "Only God makes trees!" she insisted, no doubt left in her mind that if what I had said was true I must be associated with the forces of Satan. Fundamentally, however, they were extremely practical people. My classmate's mother continued putting apples in her lunch box, and Sunday has remained our busiest sales day.

I certainly understood that in the Department of Sin and Temptation we were not in the same league as the Muse boys with their moonshine stills. Nonetheless, I've always been sure that we were somehow kindred spirits. As a child, I never knew that they too grew apples and cooked applejack. A half century would pass before I learned those stories and began to experiment with making my own apple brandy. The kinship I sensed had more to do with finding the dangerous magic in ordinary things: the white lightning in the corn kernel, the sales day on the Sabbath, the alterations in God's nursery. Just as there was fire in the cornstalk, so there was magic beneath the gold and crimson skin of that insufferably wholesome apple.

ONE

Perseverance, Pests, and Perversity

February is the month of hope and perseverance, when pruning begins in earnest and a warm spell followed by a subzero snap threatens doom. February was the month when my father would go back to the old peach orchard and clip bare branches that my mother would stick in a quart jar of water. After a week or two, my parents would know whether there would be a peach crop. If the large, plump buds burst into rosy blossoms, there was still hope for a crop, assuming we could slip past the frequent late frosts. But if the buds dried up and the branches only brought forth tiny leaves like pointy green mouse ears, then we knew there would be no July peaches. We followed this grim ritual nearly every February. Oddly enough, my mother never put the beautiful peach twigs in the Italian vases she kept for summer flowers. Perhaps the likelihood of a stillbirth from a frozen bud kept her from thinking of them in quite the same way the Chinese do when they buy forced peach blossoms to bring luck for the new year.

Forced apple blossoms, however, would go into the vases after a worrisome freeze or frost (sometimes they'd even be stolen from the orchard in full bloom to add a touch of spring to the indoors). But then apples are more reliable soldiers of the seasons. On those rare occasions when a Canadian arctic blast would plunge the Kentucky hills below the minus-thirty-degree mark, the apple buds would usually survive. Only once every fifteen or twenty years would there come a full freeze-out—and that was mostly during or after April bloom.

The effect of all this winter wind was to draw out the season of agricultural anxiety. Most field-crop growers in the deciduous zones wrestle with the weather fates from May to September. Spring brings land preparation. Autumn is the time for cleanup and equipment repairs. December to March is dead time, and more prosperous farmers, as we were during the war years, escape to Florida for a bit. But home or away, fruit growers are awarded an extra three months to worry whether the year will bring income or debt.

Our orchard was little different in its appearance from the New England orchards Henry David Thoreau wrote about in the 1860s. Full, spreading trees, their open limbs as thick as thighs, were planted in steady rows, one every forty feet. Now and again there would be a bushy crab apple tree planted to pollinate the blossoms of the "good" trees. Even in February, when the limbs hung bare and forlorn against the dead ground and the slate sky, an experienced eye could identify the varieties: Golden Delicious with their slightly bronzed bark and

fluffy form, as though they'd just returned from a Parisian coif-
feur; Rome Beauties, their long droopy limbs languishing like
one of Botticelli's squire boys; Stayman Winesaps, sturdy
trunked with short, pugnacious twigs that shone softly purple
in the bitter late-day sun; Arkansas Blacks, whose stiff, no-
nonsense arms shot straight up to the sky.

None of these varieties were planted in Thoreau's time.
With the exception of Romes, they hadn't even come into
existence. But other so-called nursery varieties, grafted onto
seedling roots, were making their way into commercial orchards. Thoreau and like-minded proponents of the wild
American prospect lamented the spread of such nursery apples
and the concomitant passing of the wild apples, those fence-
row field seedlings that bore fruit of such savage zest that it
couldn't be eaten until a hard frost had mellowed it. Our apples, most of them, were the creations of nineteenth-century
commerce, certainly the most energetic era in the history of
fruit breeding and selection. By the end of the Civil War,
American nurseries listed more than eight hundred varieties for
sale. In France, historically a world leader in apple development, there was a similar explosion in new varieties. Today,
commercial American nurseries offer orchardists about thirty
distinct varieties, and of those only about ten are sold in any
quantity. Of the four apple varieties I described above—Goldens, Romes, Staymans, and Arkansas Blacks—only Goldens
are thought viable in the global market. The others have been
replaced by newer, hardier, more disease-resistant, or prettier
apples.

Nor is it only the varieties that come and go. Today's trees
look nothing like the grand, proud fantasy trees of my or Tho-

reau's youth. While it used to take twenty-seven of those gnarly, old standard trees, each one perhaps thirty feet tall, to fill an acre, this spring we will pack nearly six hundred trees to the acre. In other, richer soils, growers may plant as many as nine hundred or a thousand trees to the acre. The difference is in the size and shape of the trees. These dwarf trees will grow to a height of only ten or twelve feet. Beneath them, black irrigation hoses will snake their way along the rows to moisten the trees' shallow roots. The old parklike orchards took ten to fifteen years to reach full production. These new so-called pedestrian orchards (because the apples are close to the ground, a ladder is seldom required to pick them) will mature in four or five years and provide a thousand bushels of apples per acre—twice the yield of the old trees.

As I write these paragraphs, I have also been evaluating the dozen or so high-tech training systems that orchardists are using to support their dwarf trees. A week ago I stood around shivering in the twenty-degree wind near Geneva, New York, along with eighty or ninety other apple growers attending the annual meeting of the International Dwarf Fruit Tree Association. Led by Cornell University specialists, we were touring several state-of-the art dwarf orchards. Some growers liked the so-called slender-spindle system, where the trees wind themselves like serpents around one-inch metal conduit braced by a stabilizing top wire. Others have switched to the Vertical Ax (for axis) system, whereby trees are allowed to grow with almost no pruning at all but have their thin, central trunks (or leaders) attached to the conduit and their bare limbs tied down like a weeping willow's. Still others support the trees' long, skinny fruit-bearing limbs by tying them with bailing twine to the central conduit until they look like rows of barren Christmas trees. In another variation, three lines of bailing twine are run horizontally from stake to stake, and the tree limbs are attached to the twine with wooden clothespins. Growers from the Pacific Northwest, where wood is cheap, seem to favor supporting their trees with horizontal high-tension wire strung between heavy ten-foot posts, but growers from Michigan and New York, where wooden posts cost double the price in Washington, prefer individual conduit stakes for each tree. Later I learned that in Pennsylvania an entrepreneur had come up with a system utilizing old carbon steel rails recast into ten-foot angle iron poles. The poles were so strong they needed no wire bracing on top, which eliminated the occasional calamity of having an entire row of interconnected trees blow over in a windstorm. But then, many growers were discovering that ten-foot wires and stakes were not tall enough to support their trees' upper canopies, where much of the best fruit grows, and it was hard to find stakes cut to eleven- or twelve-foot lengths.

Friendly arguments break out among members of the International Dwarf Fruit Tree Association, rather like doctrinal debates at an ecumenical church conference. What is beyond question, however, is the overall high-tech conversion. While these farmers are spending three or four times as much to plant their trees as they would have on conventional orchard trees, they are reducing labor costs for the future. Still more important, these cone-shaped trees will be flooded with sunlight, the critical ingredient in producing beautiful, large high-quality apples.

There is, however, a paradox in all these procedures: the

more the trees are scientifically sculpted to achieve convenience and productivity, the more closely they mimic the habits of truly wild fruit trees. The key reason that young, trellised dwarf trees bear fruit so quickly is the procreation principle. Since the tree's first obligation is to regenerate itself, it is genetically coded to make flower or fruit buds whenever it is threatened. When the lateral limbs are twisted and tied downward, the cambium on the bark is loosened, and a "threat" message is activated in the tree's chemistry. The tree reacts by putting more energy into fruit-bud creation than into leaf or wood cellulose production. If vital leaf-carrying wood is pruned, it reacts by sending up new, vigorous branches to protect the tree's respiration; if the limbs are not pruned, their growth is retarded and still more energy is devoted to fruiting. In 1991, French apple specialist J. J. Lespinasse made a further discovery while wandering among wild apple groves in France. He noticed that those trees with their centers bent over (but not broken) produced apples even more quickly. So he looked at the so-called Vertical Ax system he had developed a decade earlier and made a key alteration: in the second year after transplant—called the "second leaf"—he took the straight, central leader, bent it sideways in a broad arc, and tied it to the adjacent bracing pole or tree six feet away. Each row comes to resemble a long bushy line of great green croquet hoops. The result is that the trees bear apples a year sooner than on Lespinasse's original Vertical Ax system.

Transforming an orchard—and all it contains—into a modified, high-tech row crop is emblematic of how agriculture has been turned into an industrial enterprise subject to the vagaries of the global market. Whether the growers are Asturian cider makers in the north of Spain, wholesale producers in the south-west provinces of China (where four million acres of apple orchards have been planted since 1990), or small Appalachian farmers like ourselves, everything that we do as we pass through the growing season is being reinvented.

Consider thinning.

An apple tree and an apple eater have two very different agendas. An apple tree, like all living things, is genetically coded first to reproduce itself. To do that, it creates deeply fragrant, alluring flowers. Usually the pollen from the flowers of one variety is required to "fertilize" the flowers of another. To fertilize a Stayman Winesap, there must be pollen from another fertile variety blooming at the same time, for example, a Dolgo crab, a Grimes Golden, or a McIntosh. On any large tree, there are tens of thousands of blossoms competing for the bees' attention. Even among the flowers, however, there is an insurance system.

At the center of each cluster of blooms is a bigger, more precocious blossom that opens before the smaller buds sur-rounding it. If this "king" bud is successfully fertilized, and if it isn't nipped by a late frost, its apple will dominate the fruit formed by the surrounding flowers in the cluster, and most of the smaller apples will drop to the ground. If the king bud's apple fails or is killed, the lesser ones that bloom one to three days later will still produce fruit. But a competition will break out among them until only two or three marble-sized apples remain from the original cluster. By the second week in May in Kentucky, or June in New York, New England, or the Wenatchee Valley in Washington State, you can walk through the orchard and see thousands of tiny apples on each large tree.

A month later, half of them will be falling as their stems yellow and die.

Unattended, most of the apples on those trees will grow to no more than two or two and a half inches in diameter—containing just enough starch, protein, and carbohydrate to develop ten seeds inside each apple—for the apple tree's only agenda is to make seeds that will fall into the earth to make more trees (never mind that the seed of any apple has only a remote chance of producing a tree that will bear the same kind of apples it came from—but more about that later). So it is that a large tree with a multitude of small green apples makes a far more ambitious breeding machine than a small, carefully manicured tree with just a few big apples—which is where the apple eaters' and apple growers' agendas come in. Most apple eaters prefer to bite into large, pretty apples. To get a good supply of those, the grower strips off one half to two-thirds of all the infant apples that form at bloom time. This is called "thinning."

Until just after World War II, most thinning of excess fruit was done by men and women standing on ladders pulling the unwanted apples off by hand. It is slow, tedious work that requires a careful eye. Most people, when they are learning to thin, pull off too few apples, fearful that they are ruining the crop. The first summer my parents sent me to work in the orchard, when I was about twelve, I was told to stand on the ground and thin the lower limbs. After I had been working a single tree for about an hour, the young woman training me looked at the tree, looked at the ground, and then looked at me.

"What are you doing, feeling sorry for the little ones?" All over the ground lay the bigger apples while the smaller ones still hung on the branch. Bigger apples are easier for clumsy hands to grab, and I probably did feel sorry for the runts.

That summer was forty years ago. Families were large then and had lots of youngsters looking for summer work. Most people in our county tended their own farms, and when they'd finished setting tobacco, many were eager to get some extra money working in our orchard, even if the pay was low. Now, as we march toward the millennium, most rural residents work in towns during the day and farm on the weekends or evenings. Families have a third or a quarter as many children. There are no people to hire for jobs like thinning except migrants from Mexico, who can earn double or triple our rates by working in tobacco. The question we all ask is how to get the work done. The answer is chemistry.

Since the 1950s, research on plant hormones has produced chemicals that will induce trees to shed excess apples. The chemicals are about 80 percent effective on most apple varieties (except Golden Delicious and Empire, where the results are spotty) if used under exact temperature and humidity conditions before the fruit reaches a three-eighth-inch diameter. A week or two of hand thinning takes care of the rest. Alas, there is a problem: the most readily available chemical, carbaryl, whose tradename is Sevin, is relatively harmless to humans, but it kills *Amblyseius fallacis*, a spider mite that feeds on a tiny blood-colored pest called the European red mite. Wild trees, weighted down with knobby little apples, are protected from these red mites by an array of spider mite predators. But the orchard trees, stripped by Sevin of both their apples and their protectors, are like an immune-compromised human being: in a week or two of hot, dry weather, the European red mites suck all the chlorophyll from the trees' leaves. Along with the chlorophyll goes the trees' energy and their ability to breathe. The leaves turn from bright green to a dry, sickly bronze. The apples cease to grow, and the trees must direct all their energy to sprouting new leaves just to survive.

One solution is chemical escalation: in the early 1990s, a newer, platinum-priced chemical came on the market that promises to control mites by killing their eggs early in the season before the trees bloom. Another high-priced chemical applied just at bloom may have the thinning effect without killing the mites' predators. And dwarf trees also play a role. Because they require little ladder work, they can be hand-thinned in a fraction of the time it used to take someone standing twenty feet in the air on a ladder.

All of these maneuvers, however, have only one objective: to defeat the tree's natural determination to spread its seed. The tree accomplishes its goal by producing vast numbers of bright, small apples that cattle, deer, raccoons, and other field animals will eat from the ground after frost, digest, and then deposit as seeds surrounded by manure for the next spring's germination and growth. Indeed, a smaller apple seems to serve the seeds' purposes better, because it has less pulp surrounding them; if the seeds are not separated from the pulp before it rots at the onslaught of winter, they too will rot and die.

The business of playing with nature's design, of disrupting her fertility regimes, of recasting the very shape of her trees, seems to distress people. For some it poses a subliminal challenge to God, as with my grade-school classmate who thought we were doing the Devil's deeds by making trees with the grafting knife. I confess I too am excited, just as Thoreau was, when I happen across a messy, wild seedling hanging heavy with blotchy red fruit. Some of that excitement even infects the horticultural specialists who hike into the remote canyons of Kazakhstan and Tajikistan and western China to study the primordial apple forests that seem to have been the birthplace of the fruit. At those moments, we recognize the magic in the unadorned world of nature and are reminded of the apple's power in our most fundamental myths—from Aphrodite to Eve to King Arthur. To bite into such an apple is to bite into our origins. By contrast, to bite into a gorgeous but chalky Red or Golden Delicious from the supermarket bin is to be confirmed in our suspicions that science and the marketplace have converged to destroy nature's delights.

Here in such suspicious soil is where confused mythologies of nature are born, particularly those that pit the natural and organic against the chemical and commercial. Most of those wild trees, be they in the mountains of Kazakhstan or in our neighbors' fence rows, bear fruit that is bland, bitter, knotty, and scarred with worm tracks. Tom Burford, whose family has been growing apples since Thomas Jefferson built Monticello and who advises Monticello on its current orchards, calls most of the wild and old-time apples "quick spitters"—because they taste so bad "you spit them out quick." The good fruit, whether originally an accident of nature or a man-made hybrid, has always been the grafter's product. Greeks as early as the sixth or seventh century B.C. relied on such vegetative surgery in their orchards. The French (and likely the Persians a millennium earlier) had by the ninth century discovered a dwarfing apple tree they called the Paradise and were busily grafting other favored apples onto it. By the time the Enlightenment lit up European science, gardeners were painting their fruit trees with a variety of poisonous unctions dedicated to the death of insects. And much of what organic growers consider appropriate treatments today—copper and rotenone, to name two—are decidedly noxious to humans. Indeed, under some organic regimes, the "natural" insect poisons my father relied on in the 1930s, arsenic and lead combined as arsenate of lead, would be considered acceptable.

Very little that we eat or even plant for decoration is altogether natural, and hasn't been for some centuries. Before the Europeans crossed the Atlantic, there were no proper apples in North America; there were only thorny crabs, *Malus coronaria*, green little nubs too bitter to eat. Nor were there oranges or apricots or peaches or pears. All were gardeners' confections, formerly wild trees of Mesopotamia, central Asia, or the South Pacific, grafted and hybridized and set down in stylized plantations.

Our orchard in Kentucky sits upon a narrow strip of cleared ground called Pea Ridge on the very first foothills of the western Appalachians. To get there you drive twelve miles out from the local county seat into the mouth of a green valley through a village called Wallingford. At first to the left, and then on the right, forested ridges arise. They reach an altitude of about twelve hundred feet. At the pit of the valley, past several shanties, the road meanders to the top of "the mountain," as town people call our hill. A few farm fields have been carved out of the woods, most of them at the end of the last century. White oak, red oak, hickory, maple, ash, locust, poplar, dogwood, serviceberry, and redbud hover close to the right-of-way, their iced upper limbs sometimes forming arches over the road in winter. A half mile farther the forest recedes to an opening that is our orchard.

Off to the south, about twenty miles away, you can just see the blue outline of the next Appalachian range. If you pull onto the shoulder on an early morning in spring, kill the engine, and sit silently on the fender, it will seem as peaceful as any spot on earth. Walk farther into the orchard, place your ear to the grass or your face in the leaves, and you may be startled.

An orchard in June is a banquet. It is a raucous feast laid out in a vast vegetative ghetto ruled by a continuous struggle over death and sex. Early in May, soon after bloom, small, brown silky codling moths flutter through the dandelions and Queen Anne's lace, find mates, and lay their eggs on leaves near the flower or calyx end of the young apples, where, later, worms will flourish. Other moths lay their eggs and out pops a sturdy little worm known as the red-banded leaf roller, who pulls the leaf about himself into a tasty cocoon, which in a week will release a fresh new moth—one generation every four to six weeks. Aphids by the trillions succor themselves on the stems and leaves of grasses across the orchard floor. Then come the red mites. Their eggs have wintered over in the forest, where frost, bark, and dead leaves have sheltered them from their dire enemies, the *Amblyseius fallacis* and those familiar polychromed orange-and-black beetles known as ladybugs. In the forest only a few survive. But in the orchard they find an endless salad as they sink their long, sucking proboscises into the downy netherside of apple leaves. Finally, there are the fungi lurking in the soil, their spores ready to spew into the air on a rainy day when the humidity and the temperature both reach sixty. Then, silent and invisible, they settle over the membrane of the leaves or the skin of the young apples and form ugly lesions. A crusty scab covers the lesion, rather like athlete's foot between a gymnast's toes. Other fungal spores find their homes, bide their time, and attack the apples directly, painting brown circular rings with ridges of white dots called "bitter rot."

Sometimes, a wild tree in an abandoned field will have grown up with genetic resistance to some of its bacterial and fungal enemies; other times, nearby weeds or shrubs release odors and toxins that create shields of natural resistance, like marigolds guarding a garden. But an orchard—or even a vast field of corn or broccoli, for that matter—is an unnatural gathering of a single plant, carefully isolated from weeds and brush. The genius of organizing those plants for convenient harvest breeds special diseases. Jonathan is an apt apple example.

Crisp, red, and sweet, with an undertone of tartness, Jonathan is probably the most popular nineteenth-century American apple still widely available. It blooms late and ripens early and has long been favored by pie makers. Because it is medium in size, it fits well in lunch boxes. But it is especially susceptible to fire blight, a bacterial infection that turns the tender shoots of the tree black as though burned by fire. The disease emerges if hot, wet weather accompanies bloom; the only effective treatment is streptomycin, the same antibiotic we give our children. Infected limbs must be cut off and burned, or the infection will move down into the core of the tree and kill it. Even the pruning shears and saws must be dipped in disinfectant after each cut to prevent spreading the disease. An isolated Jonathan tree poses little threat to those varieties with greater resistance to the bacteria. But a large block of Jonathans works like an old, infectious disease ward or a kindergarten classroom: it establishes a bacterial beachhead from which the germs proliferate and mount intensive attacks on neighboring varieties like Stayman or Rome or Maiden Blush.

Red Delicious suffers a similar plight later in the summer, for Reds are a preferred specialty of the ever-hungry red mites. In just a few days, the mites turn the sweet leaves dry, along the way breeding a new generation every six to ten days, which then moves on, D-day style, in phalanxes to attack the Cortlands and the Golden Delicious and the Empires.

In some climates, where the humidity is low, the air is clear, and the temperature chilly at night, where the soils are loose, loamy, and well drained, where high mountain crests create natural barriers, the fungal and insect parasites breed more slowly. In those zones, crop oils and aerated soaps can sometimes provide a serious check to the thirty-some pests and spores that feast on the congested collection of trees called an orchard. But in the thick, soupy subtropical climates of the upper South and Midwest, and even in much of New York and Pennsylvania, unchecked parasites would soon reign supreme. To have a diversified agriculture, especially to grow high-quality fruit east of the Rockies or in Europe, means that beneath the idyllic vistas of van Gogh, Monet, and Winslow Homer there almost always will be deadly warfare between humans and bugs.

The first sprays come in March, when the trees are still

asleep, or dormant. They were an exciting, stinky operation when I was a child in the 1950s. Standing out in the front yard, I could hear the familiar rumble and grind of our old four-hundred-gallon gasoline-engine sprayer as the men filled it from one of a half-dozen small ponds and towed it through the orchard behind our powerful Case tractor. Soon a high, yellow plume would shoot up above the apple trees, a Day-Glo mist against the dark band of forest that lurked beneath the sky. Lime sulfur was what we used then, and pretty as it was, it stank like a ton of rotten egg yolks. One man would drive the tractor while the other walked among the trees dragging a long black sprayer hose connected to a twenty-inch brass spray gun, aiming it into the branches and painting them a soft yellow.

Today the dormant spray is a fine, light, odorless oil, sometimes mixed with a mite killer, or miticide. The lime sulfur used to come as foul-smelling boxes full of yellow powder; the dormant oil now arrives in fifty-gallon drums to be mixed with water in a smaller, more efficient sprayer tank that one man can operate from the tractor seat. The bright yellow plumes have been replaced with controlled horizontal mists spread by a giant fan at the back of the spray tank. The object, however, is the same as it was with lime sulfur: to smother tiny insect eggs and fungal spores that have survived the winter and will infect the tree when it awakes from dormancy.

"Dormant" is a critical, if confusing, word because it explains why apple trees were never native to Egypt or Malaysia or southern Florida. All deciduous trees enter a dormant phase. Though they may look dead, they are in fact only resting. The dormant season starts in November or December for apple trees, when daily temperatures fall below forty-five degrees and the last traces of chlorophyll have disappeared, allowing the natural red and yellow pigments to show up. Like bears in hibernation, the trees sharply reduce their rate of respiration, their breathing, while their roots continue quietly absorbing nitrogen, phosphate, magnesium, and other minerals. They awaken from dormancy when the soft Gulf breezes push the temperatures back into the fifties and sixties in late March or April. Then the flower buds and leaf buds open, but it is not the warmth alone that brings them to flower. More important still is the cold weather, which cracks the tight seal that formed the previous summer around the nascent bud. Without that descent into the cold, no amount of warm spring air can force the bud to open.

The key to battering down the season's first pests is to destroy them before they awaken. Once the petals have fallen and apples the size of BB pellets have begun to swell, upwards of a dozen kinds of worms, mites, fungi, and bacteria are ready to join the feast. The chemical warfare continues until early September, as the spraying rites are repeated every twelve to twenty-one days.

The inexorable battle moves on year after year through the Darwinian laws of survival, just as it does with human disease and antibiotic escalation. By the early years of this century, horticulturists had estimated that there were some five hundred types of insects attacking apples. By mid-century a scientific team at Cornell University had identified at least seventeen in a single family, Tortricidae—a set of moths that lay their eggs on the trees' leaves and fruits. Nearly all can be controlled by careful use of pesticides, but inevitably a few develop resistance.

One of the most famous, the red-banded leaf roller, was of no significance until DDT replaced arsenic-based sprays in the 1940s. Since then she has joined her cousin the coddling moth as a perpetual troublemaker. Following DDT, which had been so successful in clearing malarial mosquitoes but became a serious and possibly permanent threat to soil and water supplies, there came organophosphates. These chemicals, which targeted the insects' nervous systems, were a valuable improvement because, unlike DDT or arsenic, they decompose into safer compounds with little residual danger. But not only are they highly toxic when applied; the insects also develop resistance to them. As the warfare escalates, each product the chemical industry develops is more expensive than the last. Gradually, it has become clear that the potential array of pests can evolve just as rapidly as—and sometimes even faster than—the chemists' technology, and at a cost that drives more and more orchards out of business. By the mid-1990s we found that on our small five-thousand-bushel orchard in Kentucky we were spending $1,000 to $1,200 every two weeks just to keep the trees and their fruit healthy. A genuinely commercial apple grower producing fifty thousand bushels a year—which in New York, Washington, or California is a relatively small number—was paying Dow or Chevron or Merck $125,000 to $150,000 a year.

The insects' virulence and the cost of pesticides have led fruit growers to seek ways of weaning themselves from expensive chemical dependency, largely through something called Integrated Pest Management (IPM). IPM is more an approach than a method. Farmers hang small traps emitting pheromones (or sex smells) in their trees to lure amorous insects. There may be a half-dozen different pheromone traps placed in the orchard during a season. Each one provides a sample census of which insects are moving in and how close they are to mating and laying new eggs. Growers will also mount computer-based temperature and moisture sensors in the orchard to determine when the weather is right for fungi and bacteria to spread. The object is to avoid spraying until insects, mites, and fungi have reached a threatening critical mass. And some organic techniques of releasing predator mites and insects are becoming cheap enough to be commercially viable.

Every autumn our customers drive up the lane at apple season and comment on how full the trees are with fruit. They wander about the sales barn to sort through large wooden bins of apples that have been hauled in straight from the orchard. Mostly they will ask, "What's a good eating apple?" or "What's a good cooker?" Some years the Red Delicious look a little smaller, and they'll want to know why. (Usually it's the mites that got ahead of us or a frost that killed the king buds.) People want their Red Delicious to be big and pretty. If the Cortlands or the Staymans are a little smaller, they don't seem to mind. It's the taste they love in those apples. And over the years they have come to see the brownish russet on our Golden Delicious not as a blemish but as a guarantee of the sharp, nutty snap that distinguishes Appalachian Goldens from their pretty, tasteless cousins grown out West.

Often one of the men will come over while his wife is filling a bag with fruit and ask a farmer question about some trouble he's having. "I put that tree in the ground four years ago," he'll tell us, "and I ain't had a single apple on it. That nursery I got it from's just no good. Where you get your trees?"

The answer, of course, is not to blame the nursery but to ask a lot more questions. What size tree is it? A dwarf? A semidwarf? A full-size standard? Except for the dwarf, the others will take five to ten years to bear fruit while all their energy goes into building limbs. Is the soil too wet? Has the tree been staked? How has he pruned it? Too much pruning causes the tree to delay fruiting. Has he tied the limbs down or spread them apart to introduce sunlight and create fruiting buds? Each one of those questions risks confusion, especially for those old-timers who have fuzzy memories of youth, when Grandpa had a twisted old tree in the yard that just bore and bore and bore and "you didn't have to do nothin'." Chances are the memory of that old tree is like the memory of the old Ford that never had to be repaired; if Grandad were around, he'd tell another story.

These perambulations are not what our farmer-customer wants to hear when he asks why his expensive Stark Brothers Golden Delicious hasn't produced while Grandad's scruffy seedling was weighted to the ground "with the best damn apples you ever tasted." He simply wants fruit grown quickly, free of disease, and easy to pick from a convenient, pedestrian-sized tree. After all, that's what the handsome color photographs in the retail nursery catalogues promise. Since he's planted it in his back yard, he would also like it to be free of dangerous chemicals. He'd rather not spend money on complicated stakes and trellises. He would, in short, like it to be "natural."

Yet there is another sort of truth in these complaints. More than most any other fruit tree, the apple is an emblem of nature. Like Thoreau, people are struck by the wild tree's ability to survive heat and cold and produce fruit decade after decade, sometimes century after century, in almost every quarter of the temperate zone, with no help from humans. When, sometime in the Dark Ages, the early Christians adopted the apple as the tree of knowledge, likely out of a much older pagan celebration of the fruit, they set the stage for presenting the apple as nature's prince. Some classical folklorists have even traced the Greek sun god Apollo's origin to the *apfel* worshippers of the Nordic forests.

Hardly anyone except for botanical historians realizes that the apple is a creature alien to the West. As American as the apple may seem to Americans—or as Celtic as it may be to the English, or as Gallic to the French—apples are late arrivals. They were carried west from central Asia in the baskets of caravans and the guts of horses. Like any organic thing moving into a new environment, they were not resistant to all the diseases of the new territory, just as pale European colonists were allergic to the poison ivy that the North American natives used for medicine. Whereas indigenous plants—wild persimmons in North America, lingonberries in Scandinavia—evolved natural disease and pest resistances through the millennia, European and American apples have not had thousands of years to reach an accord with their hungry predators. Even if their rich genetic resources have enabled them to survive their far-flung migrations, it has taken intense human intervention to render them presentable at the tables of Caesar Augustus, Louis XIV, and Martha Stewart.

Unlike the apples we make myths of, the apples we fawn over are thus *essentially* unnatural. An apple orchard (or any orchard, for that matter) is a place out of balance with what we call the natural world. Whether we destroy the mites organically by spraying them with oils and insecticidal soaps or chemically with organophosphate poisons, we are reorganizing the natural order of things, just as we were when we cleared the land and planted the trees. To make an orchard bear good fruit, we embark on the fundamental Promethean quest: we snatch an ancient knowledge of plant survival in the wild and then manipulate it to impose a new, artificial balance on a man-made, imbalanced garden.

Each of these sophisticated advances, from pruning to insect management, has brought about an odd marriage of old-fashioned naturalism—watching nature closely—and high-tech manipulation. Even the genetic engineers who are busy mapping a half-dozen apple genomes are vitally interested in preserving the vast apple forests of central Asia, for it is in those "wild zones" that they may find the *Malus* genes that can make apples that won't rot or leaves that mites don't like or blooms that bacteria can't blacken.

Should we call these brave new orchards "natural"? The only genuinely honest answer is that nature herself is a trick, a confection of Enlightenment thinking, a projection to some lost pristine paradise that periodically resurfaces in utopian fantasies. Neither today's apple nor Grandad's was natural. Aside from stumbling across Thoreau's scattered fence-row seedlings, to find a truly wild apple we will have to march backward through the millennia, across an ocean, into the faded tunics of Mesopotamia and onto the backs of donkeys and horses across the mountains of the Caucasians and into the lands of the Uzbeks and the Kazakhs, up the streams running through the town of Almaty, called "the father of apples," where the first fruit fell to the ground long before the authors of Genesis had learned to read or write or tell the story of Paradise.

TWO

In Search of
the Primeval Apple Forest

Late one August evening in 1992, when I was living in San Francisco, I found myself scrolling through the electronic card catalogue of the University of California. The university was closed, but a laptop computer and a telephone line kept its vast cataloguing system perennially open. Obsession, insomnia, and the byzantine tracking system of the UC library had converged that summer, leading me into one of the most arcane, and yet obvious, branches of apple lore: the search for the apple's true birthplace.

I knew already that a modest disagreement had lingered for more than a century over the origins of the apple. Only a few Christian fundamentalists continued to insist that the apple came from "the Holy Lands." Waverley Root, the estimable food historian and author of *Food: An Authoritative and Visual*

History and Dictionary of the Foods of the World, declared that the first apples grew near the Baltic Sea, but he declined to elaborate on the "etymological evidence" he relied upon for his conclusion. Referring to fossils and carbonized remains of apples in prehistoric Swiss and Celtic settlements, he dismissed the conventional notion that the Romans introduced apples to northern Europe. Oddly, he seemed thoroughly unaware that geneticists have identified more than a dozen distinct species of apple, whose homes range from British Columbia to Sichuan Province, and that these prehistoric remains may bear no relation to the sweet table apples of the Romans.

Within horticulture, biologists and naturalists have pursued their own debate about the apple's origins: either southwest Asia, in the Caucasus Mountains, or south-central Asia, on the slopes of the enormous range that separates China, Kazakhstan, and Kyrgyzstan. Both areas were crossed by the great Silk Route that brought the charms of the Orient to the early cultures of the Mediterranean.

I also knew that the world-renowned Russian geneticist Nicholai Vavilov had visited both the Caucasus and Kazakhstan and favored the latter as the site of the modern edible apple's origin. But it was hard to find much of Vavilov's work in English, which was why I was spending my evenings wandering about the library's electronic nervous system. There seem to be certain unwritten, mysterious laws about electronic library catalogues. Never once did my tapped-out inquiry deliver quite the same menu of references as it had previously.

This time I asked for any title, subject, or author with the name "Vavilov" in it. The first fifteen or twenty listings were in Russian, followed by a few journal articles in English published during the 1920s (most of these concerned grains). Then came item 32. It stood apart from the rest: *The Vavilov Affair.*

What sort of "affair" could this starched-collar geneticist have been involved in? A few lines into the synopsis came another flag: "Foreword by Andrei Sakharov." What possible interest could the famous dissident and human rights activist have in a long-dead scientist?

For the next week I set everything else aside, roaming deeper into the library. I searched through old newspaper clippings and requested musty journals long since dispatched to the library's storage archives twenty miles away. Vavilov, it turned out, had traveled by mule train across southern Asia and, upon arriving in Alma-Ata, pronounced the Kazakh capital the origin of the earth's edible apples. The reason for Andrei Sakharov's interest was that Vavilov, who had identified the birth sites of more plants than anyone in history, died a grisly death in a Soviet prison, the victim of the Soviet Union's worst scientific scandal.

A few days later I learned that a small team of American agricultural researchers had opened new contacts with a group of Kazakh scientists who had quietly carried on Vavilov's work. In fact, the Americans told me, there was a remarkable octogenarian, a Kazakh native who had devoted his life to studying the vast apple forests of Kazakhstan. The man's name was Djangaliev, and his team was even now on a field expedition into the forests. Perhaps, they said, I should contact him, though the only number they had was his institute's fax number in Alma-Ata.

Two weeks later, I was on an Aeroflot jumbo jet flying from Moscow to Alma-Ata. Eighty-year-old men don't live forever, I reasoned, and apples, even wild apples, ripen only once a year. The best time to visit a wild apple forest was when the fruit was ripening on the tree.

Time is often confusing to first-time flyers in the old Soviet Union. My flight was scheduled to leave Moscow at 2 p.m. and arrive in Alma-Ata at 5:30, which it almost did, except that after three hours in the air, it was obvious we were still at maximum altitude and nowhere near the mountain range at the eastern end of Kazakhstan. No one had told me that Aeroflot listed all departures and arrivals in Moscow time—even for a two-thousand-mile flight to a city whose local time was three time zones later. At 8:30 the fat, loose-jointed jumbo jet rolled to a stop behind an attractive one-story building slightly smaller than the terminal in Lexington, Kentucky.

Dusk was losing its last light as I walked down into the plane's cargo bay and picked up my bag from the wooden luggage racks (I'd been advised that prudent Aeroflot passengers never check their bags). Inside the Intourist lounge a small greeting party awaited my arrival. A man with graying hair and expectant eyes stepped forward.

"Mr. Browning," he said, emphasizing the *g*, rather the way some New Yorkers say "Longuyland." He thrust his thick, sturdy hands out to grasp mine. Quickly his interpreter, a blond fortyish woman with matching everything—eyeliner, purse, shoes, nails—stepped up.

"I am Gallina Alexandrovna and this is Dr. Aimak Djanga-liev. We would like to welcome you to Alma-Ata." Beside her stood Kazakhstan's Deputy Minister of Ecology, the chief of the National Forest Service, another assistant, and of course the driver.

I was the first Western journalist Djangaliev had ever met.

Aimak Djangaliev was four years old when the Russian Revolution swept through the northern plains of Kazakhstan. His father, a prosperous sheepherder who could neither read nor write, had owned a large two-story house and commanded the respect of his seminomadic community until it was collectivized by the Bolsheviks. Through the eighty years Dr. Djangaliev had lived when I met him in 1992, he had survived the promises and torments of revolution, the purges and tortures of the Stalinist terror, the desperation and heroics of the Great Patriotic War against Nazi Germany, the steady, soul-numbing encroachment of the Brezhnev bureaucracy, and the final collapse of the horrible and wondrous edifice that had been the Soviet Union. Except during the war years and his time at the university, he had devoted himself to his "great passion," the study and preservation of the world's original apple forests on the slopes of Kazakhstan's Tian Shan, or Heavenly Mountains. It was in these forests, thick groves of trees that meander on for hundreds of miles, that Djangaliev escaped Stalin's agents and eventually won renown for his work. Before I arrived in Alma-Ata, my American contacts had subtly warned me to be prepared for a difficult but remarkable man whose tough ego had been the source of his salvation.

Alma-Ata (pronounced Alma a-TAH) sits between two rushing glacial streams, the Greater and the Lesser Almatinka

Rivers, that have their headwaters high in the snow-packed peaks to the southeast and disappear into a haze in the vast, arid center of the country. The name Alma-Ata, which means "father of apples," was invented by the Russians after the revolution (and changed to Almaty in 1994, three years after Kazakh independence). It had been a trading center on the Silk Route at least since the time of Alexander the Great. Imperial Russia staked out a military post there midway through the nineteenth century and gradually imported Cossack forces to hold it in the name of the czar. In those days the deep ravines and undulating slopes were blanketed by forests of apples and apricots. Green, red, yellow, rusty orange—large and sweet, small and bitter—apples ruled the land, and their forests set the acidity of the soil, made peace with specific herbs and flowers, attracted the birds and bears and antelope that would spread their seeds far and wide. That is how it appeared to Djangaliev when he was a boarding-school student there in the 1920s.

By 1992, however, when he was showing me his gardens and forests, Djangaliev could barely contain his rage and disgust at what three generations of Russian planners and bureaucrats had done to his precious fruit forest. He and his wife, Tatiana, a specialist in wild apricots, took me out to an orchard where forty years earlier he had transplanted one particularly promising strain of wild apple trees. From the smooth, tight appearance of the bark, the vigor of new growth, the size of the crop on them, these trees looked to be no more than a third their actual age. He said he had never pruned them, irrigated them, or fertilized them. The nearly ripe fruit varied in size, but much of it was as fine-skinned and colorful as New England McIntosh.

Djangaliev could see that I was struck by their appearance, and he had a good deal to tell me about these trees, but first he tugged me and his translator over to a clear space where we could see more clearly into the mountains. He began pointing with powerful gestures up to them. He was a tall man, as many Kazakhs are, and his shoulders were still sturdy and broad. The bones in his wrists were wide. His fist seemed fierce.

"If you look at the mountain over there, you'll see some dachas, some small buildings. This is one of the ways of destroying wild forests."

He was upset and wanted me, the American journalist, to know it.

"So people remove fertile land and build houses there. Well, if we treat our nature like that we'll have nothing in future, will we? The representatives of bureaucratic classes, some rich people, built their dachas in the mountains."

Now he was railing.

"They never give it a thought who produced this fragrant air and who is responsible for the beauty and fertile land. So they begin destroying it."

These were the weekend and summer cabins of the elite of Alma-Ata. Each one had a quarter or a half acre of land, tied to the highways by crude gravel drives. Already we could see hundreds of dachas; at the current construction rate there would soon be thousands. Once Western developers set to work on the pristine ski slopes just above the dacha zone, it could come to resemble Vail or Steamboat Springs.

Djangaliev is not a sentimental preservationist. He is proud of the modern industrial world the Kazakhs have built and of the productivity of modern Kazakh agriculture. He takes particular pride in the fact that, all through World War II, Kazakhstan turned itself into the breadbasket of the Soviet Union, that Kazakh grain nourished him and the troops with whom he fought on the western front. And, though he had a torturous life with the Communist Party, he retains a grain of the Marxian faith in progress and the force of history. His angst over the steady destruction of the apple forests derives from his conviction that, without such natural preserves, science and progress will be stymied.

As we stood there in the late August afternoon, he led me over to one of the wild apple trees he transplanted in 1949. The apples on this tree are medium-sized and without blemish. "This one I call *Krasota* [or Beauty]," he said, nodding his head to me, his gray eyes opening wide. That is also the name of his mother and his granddaughter, his interpreter, Gallina, explained.

This apple, Beauty, which he has studied for four decades, he hopes might become vital breeding stock in the future of Kazakh horticulture. It seems resistant to many of the standard apple diseases, offers good commercial potential, and requires no irrigation to reach moderate size. Its botanical name is *Malus niedvetskyana* number 49. A few rows away is another variety, *Malus sieversii* number 1001. This one is a large, dusky-green apple, and it grows from the tips of long willowlike limbs attached to short, stocky trunks. It may also possess special breeding qualities, he believes.

Djangaliev's counterparts in the United States and Europe are not so confident that these particular varieties will change the shape of contemporary fruit growing, nor do they consider *Malus niedvetskyana* a species distinct from *Malus sieversii*, the basic Kazakh apple. But they agree that these lower slopes of the Tian Shan and another vast, almost untouched region to the northeast of Alma-Ata called the Dzungarian Alps constitute the center of origin for the ancestors of nearly all the apples we eat today. For horticultural scientists, that is vital information. Because the apple, or *Malus*, has survived so long on these slopes, and because until recently it has been undisturbed by man, it has retained rich genetic diversity. The modern apples we find at fruit stands and supermarkets represent but a tiny slice of all the possible apples that have existed in the world. They are the descendants of thousands of years of selection for color, size, shape, and growth habits. But they are also the chance descendants of the fruit and seedlings carried by travelers of the Silk Route and wild birds and animals that ate the fruit and scattered the seed as it passed through their digestive tracts. The apples that reached Persia, Mesopotamia, the Mediterranean, and eventually central and northern Europe contain less than 15 to 20 percent of the genetic material found in these ancient Asian forests. Locked away in the genetic codes of that other 80 percent are the still unexplored possibilities of what an apple might become: Apples resistant to rots and blights and insects. Apples untouched by deep killing freezes. Apples of tantalizing yet unknown taste. Apples possessed of deep, rich skin tannins and tingling fresh fragrances that could be the basis of new untasted wines and ciders.

Even an afternoon's walk through those sun-dappled, grovelike forests reveals a variety of wild fruit that the European or American wanderer has never imagined. It is almost like a journey back into an unkempt primordial garden.

Djangaliev wanted me to see, touch, smell these forests as

quickly as possible, to absorb viscerally the intense "appleness" of this place called "father of apples." The next day he arranged for one of his expeditionary teams to pick me up for its trek into the mountains.

I was staying at a simple two-story cinder-block dormitory. A number of visiting students were also rooming there, and some older people seemed to be permanent residents. Institutions like this one had existed throughout the old Soviet Union. They were called Prophylactoriums, or preventive-care centers, and the old party elite used them as rudimentary rest retreats, rather like socialist-realist versions of Baden-Baden. This one had a resident nurse and a part-time doctor. The staff was warm, friendly. The quarters were clean. The breakfasts were greasy meat, rice or noodles, sturdy black bread, and a choice of yogurt or fermented mare's milk, a national specialty.

Once I stepped outside, apple trees seemed to be growing everywhere, along fences, between cracks in the sidewalk, and in every back yard I could see. All this in a city populated by 1.2 million people.

The team arrived by 8:30—one man, four women, the driver, and Gallina Alexandrovna. One of the women was Tatiana Nicholaevna, Djangaliev's plump, rosy wife who had spent most of her life studying wild apricots. As we drove up the roads along the mountain ravines, Tatiana pointed out how the apples and apricots occupied complementary niches. On the dry western slopes, black-trunked apricot groves dominated the landscape, many of the trees hanging on to bare rock ledges. On the moister northern and eastern slopes stood the apples.

At our first stop, across from a hill cleared for dachas, we started walking up a dirt road. Tatiana had interrupted the day's research to show me a special site. The air was clear and fresh. A quarter mile or so from the highway we turned a bend and came upon what looked like an unused children's camp. The image of a child of indeterminate sex in a pointed yellow cap and scout's uniform filled a sign.

"This is a Young Pioneers camp," she explained. "You see how they cut down the forest for these camps." Like her much older husband, she had a scolding tone to her voice. While she went on at some length about how the Party had built these Pioneer camps all over the wild forests, one of her colleagues gathered samples of wild herbs. Another plucked berries from a barberry bush for me to taste. "You can use them to flavor vodka," the colleague advised. Then Tatiana broke in, "Mr. Browning! Mr. Browning!"

Tatiana had found a twisted, ancient scrap of a tree, far beyond its natural time.

"Mr. Browning, we can call this grand, grand, grand apple tree. You see, this old tree is the mother tree. All of the daughters are much younger."

The notion that apple trees should come in mother-daughter sets was new to me. But then Gallina Alexandrovna explained that it is a matter of both language and biology. In Russian, "apple" *(yabloko)* is a feminine noun, and so the eldest in a group of trees that are interconnected through their roots constitutes the "mother" to a "family" that has the same genetic stock. The younger trees are described as "daughters." Among Russian apples, there are no fathers and sons.

An hour or so later we arrived at the expedition's fieldwork site. The highway had ended, and the dirt road dwindled to two wobbly tracks leading toward a forestry service cabin. We were nearly at the tree line, where steep blue-misted mountains are etched with giant Tian Shan pines. The few deciduous trees were small and scrawny. All this land is a state preserve available only to foresters, researchers, and the occasional backpackers who come hiking over the snow crest from Kyrgyzstan. The elevation is fifty-five hundred feet, much higher than where domesticated apples grow successfully.

The "science" the team conducted was simple. They unrolled fat orange plastic cords, which they used to mark off study plots twenty meters on the side. Then they noted down every plant in the plot, its size, health, age, and proximity to other plants. Their method was simultaneously modern and old-fashioned, reminiscent of Darwin visiting the Galápagos, scribbling down in his notebook the form and distribution of nature's creatures. For Western molecular biologists who spend their days mapping gene structures to establish species differentiation, the Djangaliev team's method is at best quaint. Some would call it hardly scientific. Yet for all its simplicity, the team's technique does produce the sort of fundamental data about how plants, animals, insects, and soil diversity work together to create successful micro-ecologies. To a degree, the very old-fashionedness of their methods might, with small technological investment from the West, lead to highly advanced work in low-chemical pest control or high-density, multi-fruit land use.

The peculiar state of Russian plant research is itself a remnant of the troubled history of Soviet science, a recurring echo of the dreadful "Vavilov affair" I'd read about in the Berkeley library. Indeed, these slopes and ravines in southeastern Kazakhstan were the site of one of Vavilov's most important expeditions.

Vavilov led botanical treks throughout Asia, South America, Europe, Canada, and the United States. He did not arrive in Alma-Ata, however, until quite late, 1929. He and his team worked their way up to the Kazakh capital with a train of mules and horses from Uzbekistan and Kyrgyzstan to the south and west. He described what he found outside Alma-Ata in his masterwork, *Five Continents*: "All around the city one could see a vast expanse of wild apples covering the foothills which formed forests. In contrast to very small wild apples in the Caucasian mountains, the Kazakh wild apples have very big fruit, and they don't vary from cultivated varieties. It was in 1929, the first of September, the time that the apples were almost ripe, one could see with his own eyes that this beautiful site was the origin of the cultivated apple."

As it happened, Aimak Djangaliev, then a teenage boarding student in Alma-Ata, met the great Russian geneticist on that first expedition into the apple forests. He was assigned to tend Vavilov's horses in the stable. He felt touched by the older man's learning and spirit of generosity. A half-dozen years later Djangaliev dedicated himself to continuing Vavilov's study into the origin of apples. Vavilov's own position, however, was becoming more and more perilous.

As Joseph Stalin tightened his grip on all the institutions of Soviet life, he grew to rely more and more on Trofim Denisovich Lysenko, his brilliant but possessed science adviser. At first Vavilov had been intrigued by Lysenko's propositions concerning genetic variation and microclimates, but closer examination showed Lysenko to be more an ambitious crackpot than a man of science. Unfortunately, the more that serious biologists rejected Lysenko as an ideologue, the more Lysenko denounced genetics as contrary to Communist thought and the more Stalin began to rely on him. At the height of the national terror surrounding Stalin's purge trials, scholars and scientists were disappearing from their posts almost daily. Vavilov began to lose his budget. His colleagues, worried for their own survival, began to shun him. A prominent member of Britain's Royal Horticultural Society, Vavilov ceased to attend international conferences. Throughout, he refused to trim his research, his papers, or his lectures to the Stalinist pattern. Eventually, he was stripped of all positions and imprisoned. In 1943 Vavilov starved to death.

Several days after I arrived in Alma-Ata and had been taken on expeditions into the apple forests, I spoke with Djangaliev in my room at the Prophylactorium and heard the story of his own life and how he had been affected by the work of Vavilov. I knew that as a university student he had attended Vavilov's lectures, but I wondered how much the young graduate student of genetics had known about the tragedy befalling Russian science.

Djangaliev sat on one side of a long table. Gallina Alexandrovna sat across from him, her back to a broad picture window. When I asked him directly what he had known of Vavilov's imprisonment, he was stopped short. Generally, he had an old man's garrulous style. His responses to questions were not so much answers as disquisitions. But to this question, he became at first silent, drumming his fingers nervously on the black-lacquer tabletop, staring out the window at the white-capped Tian Shans beyond Gallina's shoulder. His words came out slowly, suffused with sadness.

No one escaped that terror, he explained, speaking not directly of Vavilov but of his own family. His sister had been tortured, he said, forgetting for a moment the question about science and genetics and Vavilov. He had just finished his first university degree and was working at the new botanical garden in Alma-Ata when the Soviet secret police arrived one day at his sister's house. Her husband, a minor Party official, was away at work, and they began to interrogate her about his associations.

"They forced her to sit down on the neck of a champagne bottle, then pressed their weight upon her shoulders," he said. "They asked her to describe in a detailed way what kind of parties and conferences he conducted and what he talked about." As Djangaliev recounted the dreadful events of a half century ago, his eyes welled with tears, and he had to pull a handkerchief from his pocket.

"The next day," he continued, "her husband disappeared."

Desperate, his sister and her small son moved into Djangaliev's student room. Djangaliev's scientific adviser at the botanical garden realized that the young man was also in danger, given the methods of the secret police, and advised him to go directly to the mountains on a fieldwork project. It was then that Djangaliev started his collections of wild apple specimens.

He assumed that his sister would be safe. When he came back to Alma-Ata a few weeks later, however, he found his old room empty and locked. There was no trace of his sister. He went immediately to the director of his research institute for help. The director, however, understood the rules of survival under the regime and castigated Djangaliev for daring to enter his office.

He had no idea where to turn until his scientific adviser, the man who had sent him into the mountains, urged him to seek a graduate student's post. Armed only with a few borrowed rubles and a letter of recommendation from his adviser, he set out by train for Moscow, where he began his formal research into the origin and diversity of wild apples.

His adviser in Alma-Ata, he later found out, was also arrested and imprisoned. In fact, leading geneticists and agricultural scientists were being imprisoned, and executed, all over the Soviet Union. Vavilov had been removed as president of the Soviet Academy of Agriculture; eighteen of his colleagues at the Institute of Plant Breeding were arrested between 1934 and 1940. Nearly all serious agricultural publishing outlets were closed. Genetics itself, in which Russia had been a world leader for the first third of the century, was denounced as a useless, decadent bourgeois science, replaced by Lysenko's "socialist genetics." Under the rubric of "vernalization," Lysenko maintained that changing the environment under which a plant's seeds or roots were developed could "create" a new kind of plant that would pass its characteristics on to succeeding generations. Thus, grains that grew only in warm climates could,

without crossbreeding, be made to flourish in cold climates—
and thereby double, triple, or quadruple the Soviet Union's dis-
mal harvests. The theory was particularly appealing to Stalin's
own crude, socialist notions. Just as greed, violence, drunkenness, and ignorance would be erased from the "genetic heritage"
of people reared under the revolutionary regime, so a climatic
change would alter the genetic characteristics of a plant.

Lysenko's notions were little more than a warmed-over version of French botanist Jean-Baptiste Lamarck's theory of "ac-
quired characteristics," which proposed that new evolutionary
traits arose in response to environmental needs and were then
passed on to succeeding generations. Geneticists had long since
rejected Larmarckism, and nothing that Lysenko had done in
his research plots gave any credence to this resurrection of it.
But that mattered little once Lysenko had won Stalin's blessing.
By 1940, Lysenko had successfully purged any reference in
school texts to Gregor Mendel, the great nineteenth-century
Austrian father of genetics, or to Thomas Hunt Morgan, the
American geneticist who won a Nobel Prize in 1933. "Questions of Mendelism and Morganism should be removed from
the syllabus on Darwinism for the secondary schools . . . ,"
Lysenko declared, because "the secondary schools ought to be
teaching the children the foundations of science, and Mendelism and Morganism have, of course, very little to do with the
foundations of science."

"You have been arrested as an active participant in a subversive
anti-Soviet organization and a spy for foreign intelligence ser-

vices. Do you admit your guilt?" Vavilov was asked by his interrogator on August 12, 1940.

"No, I do not," Vavilov answered.

Vavilov had been arrested a week earlier on his last plant- and seed-gathering expedition into the Ukraine. It had been seven years since his previous expedition, and, according to biographer Mark Popovsky, he seemed to feel a renewed optimism. He had suffered steady attacks on his research in wheat and corn breeding, both of which were parallel to work in the United States that was sharply improving harvests. But new contacts he had developed with other powerful figures in the regime, plus approval of this trip into the Carpathian Mountains, led him to believe his fortunes were changing. All that the state Commissariat for Agriculture wanted was an evaluation of agricultural lands in the western sections of the Ukraine and Byelorussia, but for Vavilov the expedition offered a marvelous opportunity to revisit an old area of inquiry.

Early in his career, Vavilov had proposed a radical reinterpretation of where and how the world's first farming settlements emerged. Archaeologists had traditionally looked to the great river valleys—the Yellow and the Yangtze, the Tigris and the Euphrates—where the soil was rich and where they found the remains of granaries and fossilized evidence of crop growing.

Vavilov argued that farming had arisen much earlier and not in the valleys but in the mountain zones. There in the mountains, water was plentiful and conveniently available. Wild plants found survival easy. If paleobotanists wanted to find the progenitors of modern fruits and grains, he advised them to climb into the mountains.

Vavilov's own journeys into the Andes, the Rockies, the headwaters of the Nile, and the central Asian ranges led him to identify the centers of origin for scores of grains, grasses, and fruit trees. At the same time, he studied how these plants had migrated into modern farming zones. Like all Russian agronomists, he was critically interested in wheat. He knew that the early predecessor to wheat—spelt—had fed the empires of the Pharaohs and that it had come from central Asia. But had wheat arrived in Europe only from the south, through the Caucasus and on to the Mediterranean, or had there been a second route, northward through Kiev, across the Carpathians, and into modern-day Poland? He believed that it had likely followed the route of the Mongols across the Ukraine and Russia, but so far there was no evidence.

As he traveled through the Ukraine, he asked everyone he saw if they knew where any wild wheat was still growing. He asked archaeologists what foods the early Black Sea settlers ate. At a meeting with teachers at Chernovtsy University on August 5, he asked plant specialists what they knew of surviving wild grains. They knew nothing. Then, on August 6, he came across the very thing he had been looking for—wild spelt, the ancient food of Egypt and Babylon—growing unattended in the mountains. Because spelt was not native to the Carpathians and because it had long since disappeared from agricultural plantings, his discovery seemed to offer clear evidence that man's most basic food had arrived in the West by a second, northern route. To the Agricultural Commissariat such discoveries were of no interest. Indeed, they saw such inquiries as further evidence that Vavilov and his institute—which then, as now, held the world's largest collection of seeds—were enemies of the state. To waste scarce rubles on bourgeois science that could not yield better harvests immediately was tantamount to committing sabotage.

That evening, at the same time his car pulled up to the hostel in Chernovtsy, a small black Ford arrived. State agents stepped out and told Vavilov he was needed immediately in Moscow for "urgent talks." Vavilov told the doorman he would be back later.

He never returned. Instead, two other agents appeared at midnight and demanded all of the scientist's belongings, even the smallest scraps of paper, which they said he needed for his flight to Moscow. When Vavilov's colleague, F. K. Bakhteyev, attempted to enter the car for the drive to the airport, one of the agents knocked him to the ground.

"The door slammed with a bang and the car disappeared in the darkness," Bakhteyev said in a public lecture in Moscow two decades later. "It was only then that, utterly stunned, we realized at last that Nicholai Ivanovich was in real trouble."

Vavilov spent the rest of his life in prison. Over time, agents obtained "signed" confessions that he and his colleagues at the Academy of Agriculture had been part of an anti-Soviet "conspiracy." He even named co-conspirators, but all those named were already dead, and most had earned Soviet medals for their work.

Nicholai Vavilov died January 26, 1943, during the bitter cold of the siege of Leningrad, in Saratov Prison, his body wasted and swollen from starvation. Under the direction of Lysenko, a generation of brilliant agricultural scientists was under attack; many were sent to prison. The institute's research projects stopped. The curators of the seed collections lost their funding, and during the siege two died of starvation rather than consume any of the grain collection. In their place arose a generation of apparatchik students—"mutants," Vavilov called them. Writing to his son in the autumn of 1944, one of the nation's senior scientists stated, "Unfortunately all the centers of scientific work on genetics in our country have been destroyed. I can tell you by the appointment of scientists in the academy how disastrous are the consequences of that mistake on the part of the state." Serious Soviet agricultural research would not return until the 1960s and 1970s.

The great apple reserves of Alma-Ata that had so impressed Vavilov were, however, largely ignored by the Soviet agricultural elite. Because they were of no immediate use, they fell under no particular threat. Aimak Djangaliev finished graduate school in 1941 and fought on the western front against the Germans with the men he still describes as his dear American friends. It was well after the war, as rumors settled into fact, that he learned what had happened to Vavilov. After the war ended he returned to Alma-Ata and to the apple forests that had once saved his life. For fifty years he and his colleagues worked in relative obscurity with no significant outside contact until August 1989, when a team from the U.S. National Clonal Germplasm Repository mounted its initial expedition into the wild apple forests of Kazakhstan.

Researchers from Canada, Australia, and the United States have retrieved hundreds of scion cuttings and tens of thousands of seeds from the four Kazakh expeditions undertaken so far. Philip Forsline, the curator of the apple repository at Geneva, New York, and Dr. Herb Aldwinckle, a Cornell University plant pathologist at Geneva, planned most of those trips and worked with a half-dozen other geneticists and disease specialists in carrying them out. Forsline, a quiet, almost reticent fellow from Minnesota, characterizes himself as a sensible environmentalist. His principal job is the care and protection of more than thirty-seven hundred *Malus* "accessions" in the repository. Most of them, gathered from all over the world, are bitter and inedible. Many are so-called antique apples that have gone out of fashion but were once beloved by some farmer in Utica or Murfreesboro or Bismarck. Forsline freely provides scion cuttings to anyone who wants to try growing them, and in recent years, he says, one of the biggest requests has been for French and English cider-type apples (see chapter 6). His first commitment, however, is to preservation—not because the fruit tastes good or the tree is pretty but for the protection of biodiversity.

"We don't know right now what's going to become valuable or why," he told me as we walked through row after row of his first Kazakh seedlings on my initial trip to Geneva. Some of those seedlings are now ten or twelve feet high and flourishing; others have withered and died from North American diseases they couldn't fight off. Some appear to have resistance that most commercial apples lack.

Public reports of Aldwinckle and Forsline's first expedition carried a tone of bemused curiosity, generally on the order of "Real Garden of Eden Located in Kazakh Mountains." Since then, scientific and commercial-grower interest has only intensified, and more than a half-dozen research institutions have developed an array of cooperative research projects. Cornell University itself has established a cooperative research program with the Plant Genome Laboratory of the Kazakh Academy of

Sciences. There seems to be no doubt that the Tian Shan and the Dzungarian Alps hold remarkable surprises.

Expedition 4, in September 1996, proved both frustrating and deliciously tantalizing. Dissolution of the government's entrenched, hard-line authority had unleashed numerous competitive fiefdoms. Fuel shortages that had begun to surface when I was there in 1992 had grown endemic. The three-hundred-mile flight north to the Dzungaria turned into a hot, tedious overland trek by van. The Kazakh researchers themselves never knew who or what they could rely upon. One day several months earlier, they had arrived at work to find much of the tractor pool and heavy cultivation equipment gone—spirited away at night by some virtual mafia. Once the team reached the distant mountains, they were startled by their discoveries. "The surprising thing we found up there," Forsline told me later, "was that the wild apples were mostly free of insect damage compared to commercial types [in the United States]."

Resistance to scab, a fungus, or fire blight, a bacterial attack, wouldn't have been surprising. Wild apple collections will often have infected and clean trees standing adjacent to each other. But trees that could individually ward off insects are rare or unheard of. Usually, mites and insects attack bark, leaves, and fruit on all the apple trees throughout a microclimate and are regulated by other predator insects who lunch on them. But this time, one fruit tree would be attacked by codling moth worms while its neighbor would be untouched.

Tom Unruh, a U.S. Department of Agriculture entomologist at the Yakima Valley, Washington, research station, climbed up and down the Kazakh ravines with Phil Forsline.

Unruh was also fascinated by the lack of insects in the trees.

"The first thing you have to remember in a wild apple forest is that every tree is a different variety," Tom Unruh told me in a phone conversation after he returned from Kazakhstan. He wanted me to understand clearly that the conditions in a grove of wild fruit trees bore no resemblance to those in a modern orchard, even an "organic" orchard, where no chemical pesticides were in use. Because nearly every tree had sprouted from a separate seed and only a few had grown up from the roots of a "mother" tree, each tree's genetic footprint would be different. It would be a practical impossibility to produce a usable orchard that repeated the insect and disease resistance of these groves in the Dzungaria. Nor were codling moths and their larvae completely absent. About 5 to 6 percent of each tree's fruit was infested. If no codling moth had shown up, Unruh and Forsline would have supposed the moth simply did not exist there, but because the low-level infestation proved its presence, the clean trees—about one in every six hundred— were all the stranger.

"So when you have all these different 'varieties' crowded together, the fruit ripens at different times. Early summer. Late summer. Fall. You have a tree-to-tree variation. All these traits (viral, fungal, and bacterial resistance, as well as ripening time and cold and heat hardiness) segregate each one from the other. But the codling moth only has a single season.

"We grow apples out here in the West and have two codling-moth generations a year. The moths fly out in early May. They lay their eggs and hatch a new brood in July, which lays its eggs. The second generation overwinters and flies out the next spring. In California [where the winters are milder and the season longer] they have three or four generations a year. Their abundance in California makes them ten times harder to control than up here in Washington. And where we were in Kazakhstan, on the Chinese border at an altitude of twelve hundred meters, they have only one generation a year."

The mountain climate itself protected the trees, just as the arid climate in the Yakima and Wenatchee Valleys protects Washington apples from the heavy brown-rot and bitter-rot problems we face in Ohio, Kentucky, Virginia, and the Carolinas. But there was more than that.

"Remember," Unruh told me, "in a wild apple forest, every tree is a different variety. And some of them are just horrible. Inedible. Astringent. Low sugar. Woody. You wouldn't want to eat them, and neither probably would a codling-moth larva."

Whether those trees have any value to pomologists studying how to genetically engineer insect and disease resistance in tasty apples is an open question. Genetic engineers and conventional breeders have long sought a way to snatch fire blight and scab protection from varieties that have natural resistance. Scientists in New York and Minnesota and in Ottawa and Manitoba, Canada, have also crossbred eating apples with *Malus baccata* (typically, a flowering crab) from Siberia to develop cold-hardiness.

Possibly these codling moth–resistant trees have no simple utility. As far as Forsline is concerned, those present-day issues are important but not critical. Biodiversity is not, strictly speaking, a utilitarian value. The trees' unique characteristic extends the range of the apple's diverse possibilities; for the preservation of germ plasm, that is enough. Perhaps the root wood will one day prove resistant to soil fungi that don't like its taste either. Or perhaps there is something about one of these bitter apple trees that will contribute to new hard-cider varieties in which astringent skin tannin is prized and cost control favors natural pest resistance. Or the frequency of its appearance in the wild may reflect gene patterns in the species *Malus sieversii* that could open up an understanding of the apple's origins and the distribution of natural hybrids.

Or it may shed further light on Vavilov's greater quest to understand how fruits and grains migrated through the world in prehistory. One of the ongoing apple mysteries concerns the nature of the apples that appear so frequently in German and Nordic mythology, a mythology that long predates the arrival of the Romans with their sweet Mediterranean apples. The native crabs of northern Europe were surely inedible, yet the ancient Scandinavian gods and their human acolytes ate apples. What were they, and where was their origin? There is even some reason to suppose that the bitter-sharp, high-tannin varieties favored by French cider makers since the Middle Ages could be natural hybridizations of southern and northern fruit used by the Celts. Nor is everyone yet convinced that eastern Kazakhstan is the birthplace of the edible apple. A few investigators have speculated that the whole range of apples—both sweet and crab-like—existed naturally all the way from China to western Europe and that a genetic mapping of wild apples growing in the abandoned fields of upstate New York would reveal all the traits found on the slopes surrounding Alma-Ata. Should the team of geneticists and molecular biologists working with Phil Forsline find specific DNA markers that distinguish modern table apples—*Malus domestica*—from the wild Kazakh apples—*Malus sieversii*—they could lay the ground-work for an entirely new assessment of the apple's origins and the way it has been treated in myth and religion for the last two thousand years.

Just how successful this marriage of molecular biology and demographic history will be may depend more on modern urban planning than on horticultural brilliance. Perversely, perhaps, the political opening that followed the Soviet collapse has put the Kazakh apple forests at new risk. Time and prosperity seem to be working against the forests. Pristine ski slopes that were once free to Party favorites are now moneymaking enterprises and tourist attractions. Alma-Ata, a city of 1.2 million in 1992 when I visited, had no commercial sprawl either on its edge or at its center because there were only state enterprises and ragtag flea markets. Privatization of markets, however, has brought in vast global investment. In the week I was there, I learned that the Minister of Environment was negotiating a management contract with Monsanto. Chevron had won a $20 billion oil and gas concession. Two corporations, one Swiss, the other German, were petitioning for copper-mining rights, and General Electric, via the National Broadcasting Company, hoped to sew up the cable television concession for Alma-Ata. Kazakhstan has long been one of the richest of the former Soviet republics, and Western investment seems likely to generate vastly more wealth in the coming decades. Should the country follow the pattern of so many Asian nations, the "dacha sprawl" that so upset Aimak Djangaliev under the old regime may be nothing compared to what genuine, unfettered commerce promises.

THREE

The Pursuit of Paradise

"I suppose you're about to leave for God's country," Jim Cummins's recorded voice said on my Brooklyn answering machine. "I'll try you there in a couple of days." Cummins is a scientist, an apple rootstock specialist, and, in his own words, a man of prayer. The idea of God's country is, to him, deeply internal and private and, at the same time, magnificently concrete. He spent most of his life at the Cornell University research station in Geneva, New York, unraveling the genetic mysteries of God's apple trees, then breeding and refashioning them into a human reconstruction that would be more useful and beneficial to a present-day agricultural Paradise. Not long ago he and his son, Steve, opened a specialty nursery, and they were grafting several hundred experimental cider-variety trees for our orchard in Kentucky.

By "God's country," he meant our ridge top in Kentucky. Not that Kentucky was exactly his Paradise. But Cummins had soon figured out that my apple obsession was not solely about turning apples into income. I find in apples an inexplicable mystery. When I walk through the orchard, even in my most disconsolate moods, as after a freeze has taken the crop, I feel a diffuse dream-stealing magic. I wouldn't say I'm especially unusual. Most growers I've met possess a peculiar combination of Yankee practicality and quiet respect for the mysteries of the apple—be they scientific or mythological.

I found Jim Cummins through the apple germ-plasm repository in Geneva, where the Kazakh expeditions had been organized. I wanted to test a broad selection of English and French cider varieties, and the repository, with its thirty-seven hundred different apples, was about the only place to find them in North America. Other nurseries would have done the job, but Cummins, who had spent his life recasting the shape, character, and genetic health of apple trees, was intrigued. He was as curious as we were about what would happen with these difficult and often ancient varieties—some of which have been known since the late Middle Ages. If all his grafting succeeded, we would have more than forty different kinds of apples growing in our orchard. He understood that, beyond nostalgia, we would be re-creating in Kentucky our own particular God's country, our own modest Paradise in the Appalachian hills.

Paradise, of course, is the peculiar provenance of apples.

Whether among the Greeks, the Celts, the Persians, or the Christians, the apple has an ancient entanglement with Paradise, with the idea of God's country. Most modern Christians have grown up supposing that it was the apple that Eve snatched for Adam at the serpent's bidding, forever banishing them from Paradise. Although apples may have grown in Palestine at the time the biblical texts were written (Ramses II apparently received a gift shipment of apple trees from Palestine in the thirteenth century B.C., and Egyptians offered apples to the high priests, the keepers of knowledge and wisdom), no one thought to hang them on the tree of knowledge of good and evil until the fourth or fifth century A.D., when apple trees began appearing in woodcuts and ecclesiastical drawings. The Eastern Church favored figs as the forbidden fruit, while others in the Roman Church argued for the grape. Apples, however, do recur repeatedly in early visions of Paradise throughout the Indo-European world, and possession of them almost always has to do with desire, fecundity, and the reward of immortality.

In the earliest creation stories, particular gods were thought to inhabit particular plants and thereby to bring pleasure, poison, nurture, and deliverance to the creatures who ate them. To these small, short-lived, earth-scratching people, trees were possessed by the most powerful plant gods. By the arrival of the Hebrew and Greek eras, the deities had left the plant world and become sky gods who resided in a Paradise that mortals could only imagine. Even so, certain trees remained sacred and were tied to ideas of Paradise nearly into the modern era.

The word "Paradise" derives from the Persian *pairidaēza*, a contraction of *pairi* (around) and *daēza* (wall), used by Cyrus the Younger (424–401 B.C.) to describe his walled fruit gardens. Greek and Latin warriors and writers, enamored of these beautiful fruit gardens, brought them and their name west, and finally the word emerged in Middle English as *paradis*. (By the fourteenth century, a late-season, yellow dwarf apple named the Paradise was favored in French gardens and thought to be of Roman origin; a version of it now called the Malling 9, or M.9, is the leading dwarfing rootstock in North America and much of Europe.)

The Quest for the Sacred Tree

The powerful image of Paradise and its association with beauty and fecundity have not diminished in twenty-five hundred years. When Nicholai Vavilov approached the Heavenly Mountains, the Tian Shan, he found a place so beautiful, he wrote, that a traveler "might think himself in the Garden of Paradise." In his 1960 book, *Kirghizia Today*, the Russian travel writer Viktor Vitkovich repeated the story of the golden tree of Paradise: "Once, long, long ago there was an old man named Arstanbap. He had a box and in that box there grew a tree of pure gold with a nightingale singing on it. If the nightingale whistled when Arstanbap opened the box, a pistachio tree would appear. If it trilled, an almond tree or an apricot tree would rise from the ground. And if it warbled, nut trees would grow . . ." Fruits, nuts, forests filled with spectacular iridescent birds, fountains, streams, and flowers are the commonplace flora and fauna of all the Paradise stories that follow the ancient trading routes from Asia to Mesopotamia. The particular fruit and nut trees vary as the story of Paradise moves westerly, but in all of them there grows a sacred tree.

One of the oldest tree-of-life myths comes from the Babylonian epic of Gilgamesh. Traced to the third or second millennium B.C., it was recorded on clay tablets in the seventh century B.C. and parallels some of the stories in the biblical Book of Genesis. Gilgamesh, the son of a sky goddess and a demon, was born one-third mortal and two-thirds divine. A master hunter, he became a brutal tyrant once he ascended to the throne and freely ravaged his female subjects and enslaved the men. Answering the people's prayers for relief, Mother Earth created a giant named Enkidu to defeat Gilgamesh. The tyrant and the giant wrestled valiantly until finally they earned each other's respect; then they fell to their knees, embraced, kissed, and fell in love. Next the gods sent a fire-breathing monster to kill them both, but instead Enkidu held the monster while Gilgamesh slashed its neck. Finally Ishtar, goddess of desire, who had been spurned by Gilgamesh for Enkidu, sent a plague to kill the king's beloved giant.

Wracked by grief, Gilgamesh set off to find the answer to immortality. He visited his ancestor Ut-napishtim, who with his wife had survived the great flood by filling a boat with animals and plants of every sort and thereafter became the only human rewarded with immortality. Seek a magic plant that grows at the bottom of an underworld lake, Gilgamesh was told. He swam deep into the lake and brought the plant back to the surface, only to have it stolen from him by a snake. The snake ate the plant whole and thereby gained immortality for himself, each year shedding his dried-up skin to reveal a fresh and youthful body. Gilgamesh retreated home to await his death, defeated by the love goddess Ishtar.

In parts of early India, the tree of life was a golden tree. In early Persian myths, sap from the tree of life preserved the gods' immortality, while in humans it produced only drunken illusion. Usually the tree conferred both immortality and omni-science, though the Genesis account, of course, opposes the two. After eating from the forbidden tree of knowledge, Adam and Eve lose the innocence of natural ignorance and discover the self, individual, mortal, and now capable of evil. Even more threatening to Yahweh, however, their fall from innocence grants them the power of creation: through sex they can make new life. Aware of their own mortality, they were but a step away from eating of the tree of life, whose fruit would have enabled them to become gods and thereafter to challenge Yahweh's dominion.

Access to the fruit of immortality permeates all the Indo-European myths from the Middle East to the far Nordic reaches, and the farther north we go, the more that immortal fruit takes the shape of an apple. Retrace the paths of the Mongol traders, west from Alma-Ata to the Baltic and Mediterranean Seas, and the reason seems simple. Apples grow everywhere. Yellow, red, green. Large and small. Bitter and sweet. The apple is the universal fruit of the Aryan, Germanic, and Celtic worlds.

Still the apple presents a mystery that has long puzzled anthropologists, folklorists, students of ancient myth, and even contemporary genetic sleuths. How is it that Norse Odinists, Greek and Roman pagans, Judeo-Christian monotheists, and Indian Vedantics bear such remarkable inter-reflections about the sacred tree and its fruit? The gods and the myths persist even as they change their names and the trees with which they are linked. Oaks become laurels. Oranges become apples and apples become pomegranates. But as we peel apart the gods and their sacred fruits, we find shadows and antecedents that sweep across the continents and the millennia. Did the Greeks steal from the Norse or the Norse from the Romans? Was the sacred fruit of the Roman Apollo the same as the sacred fruit of the Greek Apollo, whose name, after all, is neither Greek nor Roman? Were the evil, mortal giants of Crete kinsmen to the evil giants of Germania, and did their evil schemes to steal forbidden fruit shape the much later Christian story of the sto-len fruit?

Among the Greeks the golden apples of immortality grew on a tree in an orchard hidden on Mount Atlas beyond the ocean at the end of the world, likely the Canary Islands. They were protected by the three nymph daughters of Atlas from great giants who continuously sought to steal them. Simple possession of the apples guaranteed immortality, which the gods, of course, reserved for themselves.

More than a thousand miles to the north, a parallel account seems to have coexisted in which a struggle broke out between fearsome giants and the Nordic gods in Asgard, an agricultural paradise of lakes, streams, forests, and fruited fields. There Idun, the youthful goddess of spring, was the keeper of the golden apples. Each year she served the apples at a grand banquet to preserve the gods' immortality.

Mortals, especially bands of giants, ceaselessly schemed to lure Idun away from Asgard so that they could steal immor-tality for themselves. One of the most clever giants, called Thiazi, changed himself into an eagle and swooped down on an encampment in mortal territory where several of the gods were roasting an ox. Thiazi tricked the gods into letting him have all the ox's choicest meat. Loki, a Dionysus-like god whose name signified both fire and mischief, became enraged and stabbed Thiazi with a sharpened branch. When Thiazi soared into the sky on his eagle wings, Loki's hands became cemented to the branch. Thiazi offered Loki a deal in exchange for his freedom: if Loki would smuggle Idun and her golden apples out of Asgard, Thiazi would release him from the branch.

Loki agreed. No sooner had Idun left Asgard than the gods began to wrinkle and see their impending deaths. Odin, the most powerful of all gods, who searched perpetually for knowledge and wisdom, gathered the increasingly frail graybeards into his banquet hall. Upon discovering that Loki was the last one seen with Idun, they summoned him and issued an ultimatum: he must bring Idun back, or he would become the first immortal to die.

Like the giant Thiazi, Loki was a shape-changer who could assume the form of many plants, birds, and animals, even fish. He took on the feathers of a hawk, swooped into Thiazi's fortress while the giant was away hunting, transformed Idun into a nut, and carried her back to Asgard in his claws. Thiazi, as an eagle, gave chase, but Odin, Thor, and the other gods were ready. They built a huge fire that burned Thiazi's feathers and brought him crashing to earth, where Thor crushed his head with a hammer. Loki turned Idun back into a beautiful goddess, and she fed the toothless gods small pieces of chopped apple, restoring their health and immortality.

Apples run through the surviving fragments of Irish, Icelandic, and Norse myths as restorers of either fertility or immortality. In the Icelandic Volsunga saga (a Germanic version of which forms the root of Wagner's Ring cycle), a generous goddess drops an apple into the lap of a childless king; a son is born who succeeds his father as king and plants an apple tree at the center of his court. Pre-Christian Irish myths include a divine mother, sometimes seen as Mother Earth, who both nourishes and protects the gods. A great central tree seems to have linked humans to the god-people of the otherworld that lay beyond the sea, but the nature of the tree is obscure. Five specific trees were also seen as sacred: three ash, a yew, and an oak. The oak above all had magical powers since it bore three distinct fruits each year: "apples, goodly, marvellous, and nuts, round, blood-red, and acorns, brown, ridgy." The five trees recur in the legends of the Irish kings and appear to have been valued as links to ancient wisdom and poetry. In at least one account, they were brought to Ireland by an otherworld being who carried a branch laden with nuts, apples, and acorns—all well-established symbols of fertility in Celtic and Mediterranean lore. Visitors from the land of Youth and Immortality would carry a branch of an apple tree, its blossoms silver and white, its fruit golden, its tinkling leaves so musical that mortals would be freed of pain and lulled to sleep. One description in the *Rennes Dindsenchas*, a medieval compendium of Celtic lore, describes "a shining tree like gold [that] stood upon the hill; because of its height it would reach to the clouds. In its leaves was every melody and its fruits, when the wind touched it, specked the ground."

Of all the Celtic myths and legends, none surpasses the rich concoction brewed up by Sir Thomas Malory in the fifteenth century for the story of Camelot and King Arthur. The "once and future king" had molted into medieval Christianity from Artos the bear, one of the oldest Nordic gods. As Christianity displaced the old gods, they abandoned earth for the sky, occasionally slipping back into the mortal world. Malory's version blends French and Celtic lore with the mission of the Christian Crusades. The Crusades and their attendant search for the lost Holy Grail steadily stole away Arthur's trusted friends and lieutenants until finally the aged Arthur found himself in mortal combat with his traitorous son Modred. To recover from the wounds suffered in that final battle, he repaired to Avalon, a magic "island" where the golden apples of immortality grew. Archaeological and folkloric scholars have sought out the location of the "real" Arthur, a presumed Celtic hero who fought off the Romans, likely somewhere in southwestern England. A few identify the real Avalon as Glastonbury, a town not far from Stonehenge that has carried spiritual and mystical overtones since well before the arrival of the Romans and the Christians. Glastonbury is now far inland, but before the Somerset Levels were drained, it seems likely that the town was a misty island hill permeated by deep caves—just the sort of place Arthur and his warrior knights of the Round Table would have gone to rest until the call of a new Camelot would reawaken them.

Vanity, Divinity, and Conversion

The mysteries of the apple, its rich aura of fecundity, its divine portents, its proximity to peril and immortality, raise puzzling questions about the fruit's origins. How is it that these Norse warrior gods and the first-millennium Greeks should have identified two such similar places—the island of Avalon and the isles of the Hesperides (presumably the Canaries)—beyond the ocean mists as the source of the divine fruit? Why should climates supporting such radically different agriculture and diets propose the same fruit of immortality? What sort of apples were they that both men and gods found so vital?

Part of the answer rests in language and part in the peculiar trail Christianity followed as it spread back into Europe following the collapse of Rome after the Mongol and Visigoth invasions. Save for the Muslims on the Iberian Peninsula, the great Irish and Benedictine monasteries were the only surviving repositories of classical learning and agricultural knowledge. They were principally concerned with the study and translation of the Bible. As translators wrestled with the Greek and Hebrew texts, they confronted a choice about how to tell the story of Paradise and man's loss of it. In the Hebrew text, the nature of the tree of all knowledge was left vague. In Greek texts, the word for apple, *melon*, could be rendered either as "fruit" or as "apple." The general Latin word for apple was similar, *malum*. But both *melon* and *malum* had also been used to refer to numerous fleshy fruits with seeds at their center. The fact that the words for apple and evil, *malum* and *mal*, seemed so close gave these literal-minded translators further reason to identify the apple as Eve's fruit, according to the French cultural historian Michel Pastoreau. Since for these recently converted Celts apples were by far the most common tree fruits, it required little imagination to suppose that they must have dominated Paradise. The proliferation of monasteries, with their own *pairidaēza*, or walled gardens, well planted with apple trees, particularly the small yellow Paradise apple, only further reinforced the apple's popular image as the singular fruit of Paradise.

Apples also made good utilitarian sense to a proselytizing

Christian mission. Everywhere the Roman Church has moved in the world, it has reached accommodation with existing religious icons even as it outwardly sought to annihilate them. This famed syncretism produced St. Bridget, who absorbed the Celtic Brighid (goddess of knowledge and intellect), just as it merged numerous saints with the Yoruba gods of *condomble* and Santeria in Brazil and the Caribbean. When the monks arrived in northern Europe, Pastoreau points out, they found the apple already deeply entrenched in the ancient religions: "a fruit having to do with knowledge and revelation." Given their own linguistic predispositions, they had little trouble designating it the great fruit of Paradise described in Genesis despite theological debates that lasted until the twelfth century. In effect, however, the placement of the apple in the Garden of Eden is one of the most clearly pagan acts in the development of Christianity.

Although the apple symbolized fertility and eternal life, it also bore a dangerous, even diabolical dimension. In the blending of Christian, Celtic, and Greco-Roman myths, there emerged two apples: the good apple and the bad apple, a dualism that has followed the fruit into the late twentieth century and found its origins in the ancient Greco-Roman myths. The tragedy of the Trojan War, after all, was set in motion by the rolling of the golden apple.

Zeus, who ruled over all the gods of Olympus, had designated Paris to judge a beauty contest among three goddesses, Aphrodite, Athena, and Hera, at the wedding of Peleus and

Thetis. Eris, the goddess of strife and discord, who had not been invited, tossed a golden apple at the feet of the goddesses. On it appeared the inscription TO THE FAIREST. Each of the goddesses offered Paris a bribe: from Hera came power, from Athena came wisdom, and from Aphrodite came beauty, in the form of Helen, the most beautiful woman in the world. Helen was, of course, already the wife of Menelaus, ruler of Sparta. Paris's abduction of Helen launched the Trojan War and sealed his own fate. Sundry versions of the story have been told, the most familiar in Homer's *Iliad*, but in all of them the apple seems linked both to vanity and treachery. Zeus chooses Paris to make the decision because of his good judgment in an earlier test. When he succumbs to Aphrodite's wiles, Paris becomes vain and cowardly. He prances about in his glittering armor but is terrified of combat. In the midst of battle, he sequesters himself in Helen's bedroom; finally he slays Achilles only by hiding behind an altar and stabbing the hero in the heel. When finally he is struck by a poisoned arrow, the Trojans carry him to his birthplace, Mount Ida, and beg his wife, Oenone, a mountain nymph, to gather herbs as an antidote.

"He preferred Helen to me," she answers. "Let him ask her for help." Even in death, Paris cannot escape her bitterness, for in her grief Oenone leaps upon his funeral pyre. Her spirit follows Paris into the underworld, where he must suffer eternal castigation for his faithlessness.

Glittering golden apples also brought the demise of Atalanta, nymph daughter of the king of Calydon. Like Paris, Atalanta had been left on a mountainside to die, but she was suckled and saved by a wild bear. Warned that she would die if she ever lost her virginity, the fleet-footed huntress challenges all her suitors to a race: she will marry whoever bests her, but all the losers will be slain. At last comes Hippomenes, whose patron, Aphrodite, supplies him with three golden apples that he tosses at Atalanta's feet each time she begins to pull ahead of him. As Atalanta pauses to pick them up, Hippomenes passes her. He wins the race. She is overcome by his guile. Together they run into the sacred forest of Cybele, the goddess of unrequited lust, and make love, but because Hippomenes has failed to give thanks to Aphrodite, she permits the fierce, hairy Cybele to transform the young lovers into a pair of lions.

In Greco-Roman myth, as in Judeo-Christian myth, the fruit of the gods becomes the ultimate female tool for temptation and destruction. Beneath its shiny skin lies a dangerous magic that persists today in opera, dance, and fairy tales—recall the evil witch who offers Snow White a bite of the beautiful apple. Nearly always the bewitching apple is linked to the power of love and evil.

The form and shape of the apple contain further mysteries still more appealing to the richly superstitious medieval mind and its emerging Manichaean view of the world. In an era of alchemy, numerology, and witchcraft, nothing about the physical fruit could be taken for granted. Slice the apple from stem to flower, and its female erotic imagery is plain: it is inescapably the fruit of Venus and Aphrodite. Slice it horizontally, however, and the five seeds, or pips, at its core describe the points of a perfect pentagram, a form and a number sacred at once to Christianity and to sorcery and believed to hold the key to all knowledge of good and evil. Little wonder that Merlin, the magician of the Round Table, should have chosen to sit beneath an apple tree to teach his followers and cast his spells during the transition from Celtic paganism to Christian Crusades.

Rife with magical and mysterious powers, apples were the most readily available fruit the early Europeans had. Yet even though the tree grew easily from Ireland to the Pacific coast, few people had access to the fine, sweet Roman varieties preserved behind monastery walls. The Romans had planted orchards across Gaul and Britain, and, indeed, according to food historian Maguelonne Toussaint-Samat, a pale-yellow apple from Gaul became a Roman favorite. That agricultural system, alas, collapsed along with the retreat and collapse of the empire. What was left across northern Europe was a motley collection: the native crab apples, now known as *Malus sylvestris*, a scattering of the old varieties, but mostly wild seedlings, the result of uncontrolled crosses among the Roman apples and the native crabs. Untended, the trees flourished. Often as not, however, an apple seedling will produce fruit that is small and bitter and at best suitable for cider and cooking.

As a result there was great mistrust of raw, fresh apples. Many believed them poison, just the sort of thing a witch would feed Snow White. Doctors advised against eating them raw. Even as late as the Renaissance, parents were explicitly told not to allow their children such sour fruit for fear of causing fevers and stomach flux—advice that persists into our own time about the consequences of eating green apples. At the same time, the fruit was used medicinally, ground into a pulp and applied to the skin to cure sores and painful joints, or it was pasted on the face and hair as a beauty aid—an etymological base to the skin and hair treatments we call "pomade."

Not until the monks captured the hearts and minds of the

Gallic and Frankish kings did sweet apples again become available. By the opening of the ninth century, under the rule of Charlemagne, agricultural edicts went out from the Pyrenees to the Black Forest directing peasant villages to plant a wide array of fruits, including many of the sweet apples that the monks had preserved. Even so, the old sour and bitter apples did not disappear. Most of them had been used by the peasantry to make cider, or *cydre*. Where dark, hearty ales and beers were common in the northern towns, fermented cider was a mainstay of village life. There is great debate about how and where cider arrived in France, but administrative records preserved from the Moorish rule of Spain leave no doubt that the Muslims of the eighth and ninth centuries were fully familiar with cider making (see chapter 6). By the seventh century the Merovingian king Thierry II was serving cider at grand banquets. Celtic lore suggests that some sort of cider existed deep into the past—some claim even in the Druidic era—of the British Isles, but it wasn't until the Norman Conquest in 1066 and the subsequent spread of the Cistercian monasteries that cider began to flourish there. Within two centuries, cider was well established in several English and Welsh counties.

Apollo in the Wassail Cup

It is cider lore itself that has provoked some of the most fascinating speculation on the apple's primal place in the story of Western religious tradition. If, for most of us, the image of an apple conjures up a crisp, juicy red ball into which we snap our teeth, apples for most of history have suggested a beverage. Water, we forget, was feared as a poisonous source of plague and boiling fevers. Wines were for the wealthy. Beer and cider, fermented in wooden casks through the long, dark winter, slaked the thirst of the peasants—and until the Industrial Revolution most people were peasants. Those peasants may have called themselves (or been called by their masters) Christians, but they never altogether forsook the older gods, nameless spirit deities more ancient even than Odin and Loki and Idun.

We know them too, in their cleaned-up, Christianized guises that surface at Halloween and Christmas, harvest and solstice festivals that reach back thousands of years to the time of Stonehenge. Probably the most familiar of these pagan spirits comes to us through the traditional wassail carol:

> *Here we go a-wassailing, among the leaves so green;*
> *Here we go a-wandering so fair to be seen;*
> *Love and joy come to you, and to you your wassail too,*
> *And God bless you and send you a Happy New Year.*

Of all the seasonal songs, the wassailing carol is nearly the only one left to evoke the rowdy drinking spirit of the midwinter solstice. The word itself seems to come from Saxon roots—*wæs hæil* (health be to you)—and refers to a concoction of hard cider, sugar or honey, and ale plus bits of roasted apple, all of it mulled over an open hearth. While today we think of wassail as a bracing party punch, the surviving traces of wassail custom in west and southwest England betray an older, primal ritual, an offering to the spirit of the apple tree itself. An issue of *Gentleman's Magazine* from 1746 defined "wassail" not so much as a drink but as "a drinking song, on twelfth-day eve, throwing toast toe the apple trees, in order to have a fruitful year: which seems to be a relick of a heathen sacrifice to Pomona."

Pomona was the goddess who tended the orchards of Rome's Palatine hill, but whether the English farmers had her or even the apple specifically in mind is a bit doubtful. A century earlier, in 1648, the poet Robert Herrick advised:

> *Wassaile the Trees, that they may beare*
> *You many a Plum, and many a Peare*

Later that century the naturalist and antiquarian writer John Aubrey described how in Somerset "the ploughmen have their Twelve-cake, and they go into the Ox-house to the Oxen, with the Wassell-bowl and drink to the ox with crumpled horne, that treads out the corne . . . and afterwards they goe with their Wassel-bowle into the orchard and goe about the trees to blesse them, and putt a piece of tost upon the roots in order to it." By the end of the eighteenth century in Devon, the ritual had become more standardized as the farmer and his workmen gathered with a large pitcher of cider and encircled their best-bearing tree, repeating their toast three times:

> *Here's to thee, old apple-tree,*
> *Whence thou may'st bud, and whence thou may'st blow!*
> *And whence thou may'st bear apples enow!*
> *Hats full!—caps full!*
> *Bushel-bushel-sacks full,*
> *And my pockets full too! Huzza!*

Frequently the men would fire guns through the tree, bang pans, and strike its trunk with strong sticks, raising a "cheerful noise." Then a young lad would climb up into the branches to place a cider-soaked piece of toast in a cleft, a ritual offering to the tree of its own nectar.

But then came a trick, for when the men returned to the house, they would find the doors bolted against them by the women, who steadfastly refused to let them enter until one of the men guessed what special morsel the women had put to roast upon the spit. "Some are so superstitious," one contemporary account declared, "as to believe that, if they neglect this custom, the trees will bear no apples that year."

Most of the surviving lore centers on the cider districts of Hereford and southwestern England, but a few examples of similar rituals come from Norfolk and Kent, where apples were not used for cider. Parallel stories—all of them about primal fertility ceremonies possessed by deep magic—exist in Scotland, France, and particularly the Channel Islands. British folklorist Walter Minchinton characterizes the wassailing as a primal fertility ceremony, surrounded by magic, in which the tree is a sacred phallic pillar that contains the spirit of fruitfulness "not merely . . . of apples but of all crops, of herds and of men." The dancing, the drinking, the rowdy noisemaking bring merriment to the darkness of midwinter. Even more important, the ceremony marks a generative communion. Because it takes place on Twelfth Night, two weeks after winter solstice, just as the days are growing perceptibly longer, it links man and nature, offers thanks to the aged tree for its crop, and calls on the tree to reawaken from its slumber and bring fresh fruit to the new year.

Still, as I read these accounts, I keep the old nagging question about the origins of these magic dances. Chances are there will never be any definitive answers. But surely the most beguiling speculation came early this century in a long paper from a now obscure scholar of theology and classics at Cambridge, J. Rendel Harris. Harris offered a radical, and at the time much contested, proposition: the remnants of the ancient wassailing rites betrayed not a Nordic copying of Greco-Roman traditions, as many have supposed, but just the reverse. They suggested to him that some of the Greek gods may have had their antecedents in these Nordic and Icelandic fertility spirits, most notably a northern apple god who through millennial migrations in prehistory emerged as the great sun god of Greece and Rome, Apollo.

Harris delivered his first lecture on the topic, "The Origin and Meaning of Apple Cults," at the University of Manchester in October 1917. Not too surprisingly, he drew immediate attacks from his crusty colleagues, who apparently felt that Harris had blasphemed the great Phoebus Apollo, "lord of light and healing," by reducing him to a tree spirit. Harris's lecture runs to more than forty pages; it is complex and often tenuous but rich and broad in its exploration of how relatively modern Anglo-Celtic wassail rituals seem to replicate classical fertility rites. His argument runs as follows.

As early as the fifth century B.C. images on Greek coins and vases portrayed a young female figure seated among the branches of a tree accompanied by a bird, sometimes an eagle or even a wren or a woodpecker—all of which are understood to be spirit representations of the greatest of Greek gods, Zeus. The tree usually appears to be bursting into leaf, which, classical archaeologists argue, reveals the fertility force in the tree. Another series of coins depicts an athletic boy in a tree with his hand on a feathered cock and bears the description *Zeus Felcanos*, commonly known as the boy-Zeus.

"The boy-Zeus," Harris continues, "is the proper Greek parallel to the Devonshire lad who is both tree and tom-tit [bird]," while the girl in the tree, Europa, parallels fertility rituals that persisted in the French and German countrysides well into the nineteenth century. Harris doesn't go so far as to call Zeus himself the apple spirit. He only maintains that placing a god spirit and his icons (the birds) in the tree establishes an ancient, even primordial fertility image—and, more to the point, an image, found throughout Europe and Asia (in fact, all the way to Korea), used to call forth the fertility of tree and field. Another set of fourth-century B.C. Cretan coins, he discovers, appears to parallel the boy-Zeus coins and bears the image of Apollo entwined with the limbs of a bay tree.

For a variety of reasons, Harris suspects that there were numerous other trees linked to Apollo, particularly the oak, king of the woods. Regardless, the young Apollo as a tree boy, and his association elsewhere with the ultimate fertility symbol in Greek folklore, the golden apple, leads Harris to look for Apollo's primal antecedents. The particular trees and fruits of the gods, we know well, shift as the gods fly across the various Aryan skies and landscapes. The farther north we go, the more important the apple becomes, so much so that when primitive tribes began to cultivate it they elevated the apple to parity with the oak. But, he asks, what has Apollo to do with Nordic oaks and apples?

Apollo originated, nearly everyone agrees, somewhere be-

yond Greece, possibly to the east among the Hittites, or in the
south in the region of modern Libya, or, in Harris's view, in
the far north among the people Homer describes as the Hy-
perboreans, the people "beyond the north winds" who dwell
in perfect bliss. There Professor Harris finds his most likely
candidate in the form of Balder the Beautiful, who, like Apollo,
cherished reason and moderation, and brought happiness to the
universe. Balder, son of Odin and Frigg, was the most desirable, most beloved of all the Nordic gods. His radiance lit the
lives of all around him, and he lived in a golden palace. To
protect him, Balder's mother, Frigg, drew promises from every
object—men, animals, water, rocks, wood—never to harm
him. All but one agreed, and that was lowly mistletoe, whom
Balder's mother considered too insignificant to ask. At grand
banquets, humans and gods alike grew rowdy, and Balder de-
lighted them as they banged and struck him with sticks, rocks,
and spears—all to no effect. (Recall the wassailers banging on
the apple tree.) All the universe would have persisted in eternal
amity had not Loki, the god of strife, approached Balder's blind
half brother, given him a mistletoe-tipped arrow, and helped
him release the arrow straight into Balder's heart, after which
the universe descended into darkness.

Fair enough to see a dark reflection between Balder and
Apollo, even if Apollo never perishes. For Balder's linkage to
the apple tree, Harris turns to archaic roots of the word for
"apple." In old Norse, "apple" is *abal*, with its accent on the
second syllable. Old English renders apple tree variously as *ap-
pledore* or *apuldre*. Harris wonders whether such references, including *apel-dur*, pronounced BAL-DUR, may lead us back
to Balder. "According to these suggestions," he concludes,

"Balder is the apple-boy [preceding Apollo as the tree-boy at Delphi], . . . the oath taken by Frigg [presages] the spells for the good luck of the apple-tree, and the beating of Balder with sticks and stones [is] a part of the rustic ritual . . ."

Fertility and Mountain Folklore

To our modern, rational minds, it surely seems a lengthy stretch from Delphi to the Icelandic sagas to "Here We Go a-Wassailing." Few contemporary scholars would approach the source of food myths in quite the same way as Rendel Harris, nor would they assume so easily that the Nordic myths and stories necessarily preceded those from Greece and Rome. How can we be sure, for example, that Balder and the whole Odinist panorama were not degraded versions of the Greek empyrean, as rendered through centuries of Viking and Saxon expeditions to the Bosporus? Just as troublesome is the fact that most surviving versions of the stories have been refracted through the proselytizing eyes of Christian monks.

And yet, the apple, its perfect pentagram hiding beneath a blushing skin of red and gold, continues to lure us with a wider array of myths and rituals than any other fruit we know. Whether or not its tree spirits gave birth to the Norse and Greek gods of healing and light, apples persisted through the Middle Ages and the Renaissance into contemporary times as powerful fertility symbols. Paintings and sculptures unearthed in the nineteenth century demonstrated conclusively that apples were the fruit of Aphrodite and Venus. Aristophanes, Theocritus, and a handful of other Greek dramatists regularly fashioned comic punch lines around the resemblance between apples and a woman's breast, a visual refrain that has marked the entire sweep of Western art—from the high Renaissance paintings of Barthel Beham and Lucas Cranach in which the Virgin holds an apple beside her breast before the nursing Christ child, to the succulent Impressionist apples of Cézanne's still lifes, to the blatantly seductive apples of Picasso and numerous twentieth-century modernists. Across the millennia and across the entire sweep of the Eurasian landmass, apples have remained the quintessential fertility fruit.

From the coast of Brittany to the Sea of Japan, these half-forgotten and re-remembered fragments of folklore still follow the apple. Even in this century, rhymes, jingles, and traditional predictions of love, marriage, and fate persist. A young Sicilian girl will toss an apple into the street beneath her window to find her fate; if a boy picks it up, they will marry within a year; if a woman picks it up, she must wait another year to try the toss again; and if a priest picks it up, she will die a virgin. Traditional Kyrgyzstani women who are unable to conceive will roll themselves on the grass beneath an apple tree to make themselves fertile. A newly married Montenegrin woman will throw an apple against her husband's house to encourage the birth of many children. In provincial towns of northern France, a young woman will twirl an apple peel three times around her head and throw it into the air, and when it falls it will form the first letter of her true love's name. In the Walloon country of Belgium, a young woman tests her fate with apple seeds: she places the seeds on the lid of a hot pan and asks, "Will I like him?" If the answer is yes, the seeds will explode.

The game continues through a succession of questions—"Will he marry me?" "Will my first child be a boy?"—and each time the answer is yes if the seeds pop open. Finally, she asks how many children she will have, and the answer is the number of seeds that have burst.

Many of these folkloric customs crossed the Atlantic with the immigrants who settled America, and because apple trees have covered farmland in nearly all of the eastern half of the country, the myths have flourished as local superstition—nowhere more than in the hills neighboring our farm in Kentucky. When I was a child growing up on the orchard, I would hear my father ribbing young couples about how important it was for them to eat more apples because they were "high in fertility vitamins." Such fertility vitamins have yet to be found, but his banter, born of his classical training in Latin, may well have resonated with our country customers. "Eat a crab apple without frowning and you'll win the love of your dreams," runs one Kentucky mountain saying. "If you can pull an apple apart with your bare hands, you can have any girl you want," goes another. Folklorist Daniel Thomas recovered a treasure trove of Kentucky apple lore when he hiked around the Appalachian hills early in the century—and, like the European versions, most of the stories used apple seeds to predict love and childbearing. Stick apple seeds to your forehead, and the number that stay will tell you the number of days left before you see your sweetheart. Collect fresh seeds, give them names, and flip them in the air; the seed that strikes the ceiling names the one who loves you best. Press five seeds with names to your face, and the first to fall will be the one you marry.

The seeds, so precisely placed at the points of the hidden

pentagram within the apple, account most for the superstition so long attributed to the fruit. Apples are prolific bearers of seeds, and indeed in French and other Romance languages the word for seed and semen remained the same until well into the nineteenth century. To identify yourself and to surround your family with such seed were closely akin to making an offering to the gods and goddesses of fertility, and thereby to guaranteeing your family's future. Nearly all common fruits, grains, and flowers—quinces, corn, wheat, walnuts—have at some time or another served as talismans of fertility. None, however, has equaled the mystique of the apple. Not until our own century, however, would geneticists and molecular biologists working at the frontiers of biotechnology understand what it is that has made the apple the most fertile of fruits.

FOUR

Genetic Promiscuity, Biotech, and Suicide

In the beginning there were roses.

Small flowers of five white petals opened on low, thorny stems, scattered across the earth in the pastures of the dinosaurs, about eighty million years ago. Temperate, forgiving climates blanketed most of the Eurasian landmass, and gradually North and South America pulled away, opening the wide Atlantic Ocean. The primal roses moved with the land, and as the continents grew apart, the roses crept forward through the millennia, each on its own distinct evolutionary trail. These bitter-fruited bushes, among the first flowering plants on earth, emerged as the vast Rosaceae family and from them came most of the fruits human beings eat today: apples, pears, plums, quinces, even peaches, cherries, strawberries, raspberries, and blackberries. Of all the fruits, however, the apple was the most genetically notorious.

The apple, as some botanists have described it, was the unlikely bastard child of an extra-conjugal affair between a primitive plum from the rose family and a wayward flower with white and yellow blossoms of the Spiraea family called meadowsweet. How the mating happened is unknown; all that is left of either ancient plant are fossil remains trapped in peat. But paleobotanists believe that eight pairs of chromosomes from the plum were united with nine from the prolific meadowsweet, giving birth to the first form of *Malus*. The offspring bore no resemblance to our honey-sweet Fujis, our spicy Cox's Orange Pippins, or our tart, juicy Granny Smiths. These crude "apples" were as tiny and bitter as rose hips and almost certainly toxic. Yet within their thirty-four chromosomes, there resided the genetic material that would grow into the hardiest, most resilient, and most diverse fruit on the earth; there would eventually be more than ten thousand varieties.

Cyanide, one of the deadliest poisons in nature, was the key. Molecules of cyanide contained in primitive apple seeds bonded to plant enzymes that were vital to the process of photosynthesis, which the tree uses to breathe and grow. Although the process is not entirely clear, the cyanide seems to have helped to regulate the growth of the wild trees, enabling them to adapt to an enormous range of soil and climate conditions—from the gardens of the Pharaohs to the Siberian tundra. While there is some debate among paleobotanists over the original hybridization that created the apple, its complex genetic structure—normally thirty-four chromosomes, though occasionally fifty-one or even sixty-eight triploid and tetraploid varieties—accounts for its radical promiscuity. Apples have been crossbreeding for so many millions of years that the parentage within each seed contains a dizzying mélange of inherited characteristics. These complexes of dominant and recessive genes create such an extensive chain of DNA molecules that any one "mother" tree can produce a broad array of similar-looking apples whose seeds will produce "daughter" apple trees that have completely different shapes, resist or succumb to different fungal, bacterial, and viral diseases, and create fruit with utterly different color, sweetness, hardiness, and shape.

Left to themselves, as the European orchards were after the collapse of Rome, apples will regenerate their own enormous library of genetic information. A handful of varieties, crossbred by wild bees, insects, and the wind itself, will create scores, and eventually hundreds, of new varieties that quickly drift away from the profile of their parents. A hearty number of those that survive will backcross with their own ancestors. Most die of disease, drought, cold, heat, or damage done by grazing animals, but the majority of those that do survive untended are, as the medieval peasants knew, not likely to produce sweet, large, fragrant fruit. These sturdy members of the genus *Malus* seem to contain enzymes and other chemical packages that ward off potential attackers and that carefully nurture the progeny within—the seeds—which must survive to make another tree. They are small, bitter, colored red, green, and yellow, exactly like most of the fruit that Nicholai Vavilov and his Cornell University successors found as they hiked the forests of Kazakhstan. Small, bitter apples that resemble much of the Kazakh fruit also cover the hillsides of Basque and Asturian farms on the north coast of Spain. Local farmers there told me in 1996 that they grow upwards of eight hundred varieties for their local cider presses, nearly all of them probably descendants of the orchards brought to Iberia by the Romans two thousand years ago. Yet there is a mystery locked inside these old trees. If they are simply the result of feral evolution from Rome, and therefore from mother Kazakhstan, then the DNA codes locked within them should betray their Asian ancestry. If, on the other hand, they contain gene sequences that are not found in apples along the great Tian Shan range, then their ancestry raises a raft of new questions about the origins of the European apple (and perhaps about the ancestry of the ancient European faiths that were so heavily entwined with apples).

Before the Europeans arrived, Geneva, New York, was called Kanadesaga, chief village of the Seneca nation. The Seneca had planted large apple and peach orchards, likely from seeds given them by Jesuit missionaries. General John Sullivan attacked Kanadesaga in 1779, burned the village houses, and girdled the apple trees; in his relentless campaign across western New York, Sullivan destroyed forty villages and their orchards. The state of New York established its Agricultural Experiment Station at Geneva a century later, in 1880, and merged it with Cornell in 1923. In the 1980s, the USDA built its apple and grape germ-plasm repository. Together, the Cornell and USDA installations have won widespread recognition as the world's leading center for apple investigation, rather like a pomological National Institutes of Health (NIH). Breeders, preservationists, disease pathologists, genetic engineers, and food chemists have formed an easy cooperative of applied research that has "created" more than seventy distinct apple varieties. I've made a half dozen trips to Geneva during the last decade, mostly to keep track of the Kazakhstan research and to check on how the wild Kazakh seedlings were progressing. Eventually I hope to see how they will fare in Kentucky mountain ground.

Stan Hokanson made his first trip to Kazakhstan in September 1996 with Phil Forsline, the director of the apple repository. It was the fourth expedition that the USDA and Cornell had mounted. Seedlings from the first trip, in 1989, had already grown ten feet tall at the repository orchard a half mile from his lab in Geneva. He had come to Geneva fresh from graduate school at Michigan State University expressly because of these expeditions to Kazakhstan. One November afternoon in 1992 he was driving home when he heard my National Public Radio report about the Kazakh apple program.

"I looked at my wife and told her, 'When I finish my Ph.D., I'm going to write those folks and ask for a post-doc.' " When they received his proposal, the USDA scientists at the Geneva Experiment Station had no money for post-doctoral studies. But Hokanson made a beguiling offer. He wanted to develop a genetic taxonomy for the repository's apple holdings by applying the new biotechnology analysis he had mastered at Michigan State. In his doctoral work he had focused on issues of plant diversity and evolutionary biology, but nothing caught his fancy quite so much as the chance to work with these Kazakh apple forests. "I knew I wanted to look at the genetic diversity that existed in those wild Kazakh forests and at the genetic diversity there in the Geneva repository population. If there was a lot more diversity in the wild population than we had in our collection, then that would be a big problem." As a young scientist, Hokanson could further benefit from the cutting-edge work of Cornell geneticist Norm Weeden two buildings away. Weeden has devoted most of his professional career to an international "apple genome" project that will eventually produce a complete genetic map of the principal apple varieties.

Preservation of genetic diversity in grapes and apples is the repository's central objective. Besides collecting its three thousand–plus varieties in the field, it had developed a technique to back up the collection with both seeds and bud wood held in liquid-nitrogen containers at the USDA cryogenic center in Fort Collins, Colorado. Even so, researchers had only a general hunch as to how much they had captured of the huge genetic range that existed in the Tian Shan. Genetic engineers had spent years mapping the genetic codes of several field crops—tobacco, corn, and other grains—but scientists at Cornell and in Europe had only begun their first apple genome projects. No one knew exactly what lay in the living genetic library of the repository's holdings. Until they knew what they already had, they couldn't know how much was yet to be gathered.

There are two ways to manage genetic preservation: save the wild site, in this case the Kazakh forests, where some trees are as much as three hundred years old; or bring seeds and tissue samples back to repositories like those in Geneva. "Ideally, we like to do both," Hokanson said, "but there's a problem bringing them here. When we bring them back, they're frozen in time. They won't continue to age, evolve, and be subjected to all the challenges of nature that they would in the Kazakh forests. That's why it's so important to preserve as much of those forests as we can. It'll be lost for perpetuity if the forests disappear. Once they're gone, they're gone forever."

Hokanson's first assignment was to develop genetic markers for the repository's "core collection" of sixty-six varieties. Much the way crime labs identify individual blood samples and AIDS researchers track strains of HIV through the use of polymerase chain reactions, or PCRs, plant geneticists can mark and identify groups or sequences of pairs of molecules that bond the two strands of DNA in plant tissue. These molecular patterns, labeled, say, with letters like At, At, At, At, or cG, cG, cG, recur in differing lengths at varying positions within the gene and can be measured by a gene-marking technique called SSR, or simple sequence repeats. Using SSR and similar techniques, geneticists can distinguish one gene from another by measuring the position and length of cell repetitions within the genes and then correlate particular genes with particular traits in the organism. In humans these gene sequences can signal a genetic predisposition toward diseases like Huntington's chorea or myotonic dysentery. Plant scientists like Hokanson and Weeden use the sequences to identify distinctive genetic groups.

To begin, Hokanson wanted to code the most common apple groups used in modern orcharding and plant breeding. That stage was fairly easy. First were the Macs—Jonamac, Macoun, PaulaRed, Empire, among others—which owed half or more of their ancestry to McIntosh; then the Goldens—Jonagold, Mutsu, Firmgold, Shinsei, Smoothee, and so on—which derived half or more of their ancestry from Golden Delicious; then a Jonathan family and a Red Delicious family. Those groups were easy to tag because the parent apples have been used as primary breeding stock since the last century.

As Hokanson worked his way through the core collection, he stumbled across what he calls "a few titillating findings"—gene codings that seemed to identify the murky ancestry of old commercial apples. One of those, called a James Grieve, was introduced at the end of the last century by Scottish nurseryman James Grieve, who distributed it to growers in Denmark, Holland, Sweden, and Germany. The James Grieve's physical characteristics—its taste, shape, hardness—and the tree's growing traits indicated that it could have been either a seedling of Pott's Seedling (which had been discovered growing in a gooseberry hedge around 1850) or a seedling descendant of the favorite English apple, Cox's Orange Pippin. After running an SSR gene-sequencing analysis on all three, Hokanson was able to show that the James Grieve fell much closer to Cox's Orange Pippin than to Pott's Seedling. Identifying the Grieve's ancestry may seem like an arcane sidetrack in the fast-paced world of biotech but for the fact that at least eight variations of the Grieve—some darker red and later ripening, some brighter and earlier, some that produce bigger crops—have been identified and introduced this century, one as recently as 1960. Because the Grieve still has a popular following in northern Europe, breeders might want to use it in their search for newer, higher-quality apples. Even a crude genetic outline of the Grieve's parentage may enable them to foresee strengths and weaknesses in new crosses, or through genetic engineering to introduce resistance to the diseases that have long plagued Cox's Orange Pippin, or even to identify specific gene complexes that are desirable in the Cox-Grieve lineage and introduce them into new laboratory crosses.

Once the repository's entire core collection has been genotyped, or genetically fingerprinted, the Geneva researchers can
create a phenogram, a sort of genealogical chart for each apple
that shows the size and shape of its family tree, how it is related
to other apple families, which characteristics seem to persist
within and across family lines and which seem to disappear.
Chances are, however, that only a small percentage of the total
universe of apple genes will show up on that phenogram. Even
trees grown from Kazakh apple seeds seem to contain no more
than a fraction of the genes that exist in the Kazakh forests,
which is why the 1996 Cornell-USDA expedition team
brought back leaf samples from many of the older trees that
appeared to have high-quality apples and high disease resistance. Only after those living trees have been genotyped will
the gene mappers know how much is at risk and how much
must be saved as modern development begins to encroach on
those forest reserves.

The more I learned about the Geneva genotyping project,
the more I wondered whether this high-tech pursuit of pomological pedigrees could settle my questions about the origins
of European cider apples. Hokanson was intrigued, as were
others at the Geneva complex, with the questions, but it soon
became clear that the answers would not come easily. At the
time, May 1997, Hokanson was ready to move toward the
project's next step: using gene-sequencing markers to charac-
terize an entire species of apple. Just as certain recurring sequences could identify smaller variety groups like Delicious or
McIntosh, so, logically, should a larger set of distinctive genetic
patterns identify a single apple species. The trouble is, what
makes a species? Simple as that question might seem, species
designations are the source of great debate among taxonomists,

the people who take it upon themselves to differentiate and label the plants and animals of the world. Since the great Swedish botanist Carolus Linnaeus laid the foundation of modern biological taxonomy two and a half centuries ago (in *Systema Naturae*), botanists have used, rejected, revised, and resurrected terms like species, form, race, type, and family—and that's just for the world of plants.

Broadly speaking, there are "splitters" and there are "lumpers." Russians historically have been splitters. The British and Americans have tended to be lumpers. Where Americans would find two or three species of plant, Russians, with their predilection for intricacy, would identify six or eight. Among apples the Russians have argued that there are nearly eighty distinct species, twenty to be found in the former Soviet Union alone. Once, as I listened to a translated discussion between several Cornell scientists and Aimak Djangaliev, I gradually began to realize that the Kazakh geneticist was growing uncommonly cantankerous. The subject was a group of apples with red bark, red leaves, red cambium, and red flesh that the Russians and the Kazakhs label *Malus niedvetskyana*. Almost none of the Americans or West Europeans consider it anything more than a variation of the large general species found throughout the region called *Malus sieversii*. In the face of such arguments Djangaliev developed so fierce a defense of *niedvetskyana* that the question of its taxonomic existence seemed to carry the color of personal or even national pride.

Until recently, most pomologists agreed that there were about thirty-five different species of apples in the world, most of them knobby, indigestible little crab types that one could well imagine had sprung from the chlorophylled loins of a rosy prune and a promiscuous wildflower. The four species indigenous to North America are all crabs: *Malus fusca* (mostly in Alaska and the Pacific Northwest), *Malus angustifolia* (mostly in the Southeast), *Malus ioensis* (mostly in the center of the country), and *Malus coronaria*, a long-thorned tree, one of which grew at the edge of our yard in Kentucky, burst into dark-crimson flower in late April, and bore sour green fruit the size of olives late in August. *Coronaria* proliferated throughout the eastern two-thirds of the United States. *Malus baccata*, from Siberia, was used in the last century to breed cold-hardiness into regular eating apples, and one of the crabs, *Malus floribunda*, has proved useful for introducing disease resistance. However, there seem to be only three species that humans have cared much about: *Malus domestica, Malus sylvestris*, and *Malus sieversii. Domestica*, as its name suggests, is our basic everyday apple and is likely a blend of two or three species. The largely green and yellow crabs of northern Europe are *Malus sylvestris*. The Kazakh apples, which appear to be the great mother fruit, are *Malus sieversii*.

Western scientists have assumed that our familiar *domestica* is descended primarily from two wild Eurasian species, *sylvestris* and *Malus pumila* (now regarded as apocryphal). But what about the bitter-sweet and bitter-sharp apples used for English and European cider making? Should they turn out to be simple wildings of *domestica*, with slight traces of *sylvestris*, then there are no surprises. Should it turn out, however, that these cider apples are predominantly of *Malus sylvestris* stock, then there's much more reason to believe that there was an ancient apple-eating population in the north country long before the Romans arrived bearing their carefully grafted apples. The presence of a sharp but still edible and indigenous northern apple could also lend some credence to Rendel Harris's proposition that the English and Norse offerings to the apple spirit have gen-uinely ancient roots and could have long preceded the Medi-terranean apple myths and divinities. Perhaps most important, such genotyping of these old apples might provide new insights into the overland routes that apple-eating Aryan tribes took on their way to Europe.

Chisel Jersey, a bitter-sweet cider apple from the Somerset region of England, may open the door to that set of mysteries. After Stan Hokanson entered the genetic data for all sixty-six apples in the Geneva core collection, Chisel Jersey fell to the bottom of the chart. "Chisel Jersey is the most diverse thing we have in that collection, the apple that's the most *unrelated* to all the others, though for now we still consider it *domestica*." In short, it shared fewer genetic markers with the others than anything else in the collection. The question, however, is whether it shares distinctive genetic traits with another species, such as *sylvestris*. The answer to that question, however, is vastly more difficult. Hokanson's next job was mapping 142 non-*domestica* varieties in the core collection to find family pat-terns among them. Only then can molecular biologists load data from the two groups—comprising 208 varieties—into their computers and hope that the software is capable of grind-ing out the correlations that will begin to reveal the ancestry of these peculiar cider fruit.

For thirty years Roger Way worked in a great, glass greenhouse at the Geneva experiment station. Each spring he would walk out into one of the Cornell orchards and carefully snip apple blossoms just as they were forming creamy pink balloons. Then he would bring them into the greenhouse, pull back the petals, and gently run a comb up the fragile filaments called stamens, severing the anthers, the yellow pouches at the top. The anthers would fall into a petri dish, and he would set the dish in a warm, sunny spot for several days.

If he had chosen them at the right moment, if he hadn't bruised them during the combing, if the warmth of the sun was just right, the anthers then would open and free the grains of pollen guarded inside. If his timing was wrong or his technique too rough, the anthers would wither. Once the pollen was ready, he and his assistants would return to the orchard. First, using their fingernails, they would carefully pinch the outer base of the unopened, "balloon stage" flower and remove the petals, the sepal, the stamens—everything except the central stigma and the ovaries that lay at its base, thereby eliminating anything that could attract bees and foreign pollen. Then, they would touch a fingertip into the collected "male" pollen and dust the "female" stigmas with it. That way they could be sure that only the correct, selected pollen was fertilizing the flower's ovaries.

Roger Way came to Cornell as an apple breeder in 1949. He ran the breeding program from 1962 through the 1980s. "In my thirty years," he told me, "I probably went through seven hundred thousand apple seedlings. Only now and then one clicks." Two of those seedlings that did click became world famous: a large, crisp red-gold apple called Jonagold and a medium-sized, sweet ruby-red apple called Empire. In western Europe growers are planting more Jonagolds than any other apple, and in the Midwestern United States Empire is running strong competition to McIntosh and Red Delicious.

Jonagold was a classic case in fruit breeding. The initial cross, of Jonathan pollen onto Golden Delicious stigmas, was executed in 1943—six years before Roger Way arrived at Cornell—but it wasn't until 1953 that the apple was selected. And it wasn't until 1968—a quarter century after the initial cross—that Cornell released Jonagold to nurseries for commercial sales. Traditional apple breeding is a slow, often exasperating process.

First came the original pollination in the spring of 1943, which produced hundreds of apples. All of those apples were Golden Delicious. That fall they were picked and the seeds dug out. The seeds, not the apple, are the children of the cross. The breeders kept about six hundred good-quality, well-formed seeds, chilled them to forty degrees, and held them in a damp medium for at least three months. Each seed was planted into its own pot the following spring, but only about half grew into seedlings. Of course, the Golden-Jonathan was only one cross the breeders were making. That year, as every year, there were also several other varieties being crossed, making thousands of pots of apple seedlings to be tended. After a month in the greenhouse the seedlings were transplanted to nursery rows, where over the summer they grew to two or three feet in height. The healthiest found their way to a test orchard in the spring of 1945, where they remained for thirteen years. Twenty percent of them never bore fruit. Many died of bacterial and fungal diseases along the way. Most that did fruit bore small apples that were either bland or bitter.

Roger Way had been retired from Cornell for fourteen years when he came in to meet me at the Geneva research station. Until a year or so earlier he had kept an office there and would often show up for the daily 10 a.m. coffee klatch. A rather formal man, he arrived in a tie and jacket with two fat files of manila folders under his arm, retrieved from the second-floor archives. Those archives contained notations for all the fruit-breeding work Cornell has carried out since the 1890s—easily the most extensive apple-breeding program in North America.

"When we picked this one out, its original designation was NY 43013-1. Forty-three for the year the cross was made. Thirteen is the thirteenth cross in that parent group [Golden Delicious and Jonathan] made in that year. The dash-1 indicates it was the first selection we made of that parent group. Next would be dash-2 from that parent group." Breeding apples, he reminded me, is a lot more like breeding people than peas. In Gregor Mendel's famous pea-breeding research, each cross between the same two parents produced the same identical, or near identical, offspring, meaning that peas are homozygous. Apples, like humans, are heterozygous. "Each seedling is a sister seedling because they have the same two parents, but like a human family, every seedling is genetically different because of the highly heterozygous nature of the apple."

We riffled through the files for half an hour and then returned to the archives to pull down another set of yellowing cards and lists typed out on onionskin. Everything was perfectly preserved. Transplant records for 1946 were on a full-page chart listing two dozen crosses under study: Bedford × McIntosh, Macoun × Northern Spy, Golden Delicious × Bedford, Grimes Golden × McIntosh, Red Spy × Champ Reinette, Red Spy × Wagener, and many others. All were planted November 2, including fifty-two Golden Delicious × Jonathan, in row 20 in Orchard 1. Turning to the field files for Orchard 1, we looked up the breeders' notes for row 20. Numbers 54 through 105 covered the Golden × Jonathan seedlings. The first recorded apples on those trees were taken October 2, 1951, marked medium-sized, oblate, bland, and acid, followed by a large *D* for discard. Another batch, taken October 6, 1954, was round, 90 percent red-striped with yellow flesh, "good quality" but "poor texture—Discard." Nearly all the apples on trees 54 through 67 were small, red, and followed by a *D*.

Tree 69 gave the breeders their first break. Notes for October 12, 1953, described a large, red conic apple with yellow flesh "good—juicy—has Jonathan-like flavor." Off to the side of number 69, Roger Way had written "*Prop*/53," which told the technicians to take bud wood from the tree and graft it onto dwarfing rootstock for further study. Notes from 1954 and 1955 on the original tree continued to show a pretty, juicy crisp apple. Not one of the other forty-two seedlings—ten had died—showed even a hint of promise; all carried the damning *D*.

I asked Roger if he remembered how soon he began to get excited about this one promising cross. His answer was stoic, modest: "Well, you see, we've been burnt so often we never arrive at that point until the growers start telling us it's a good variety. We think it's going to go, so we put a name on it, but we don't really know it's going to go until the growers come back and tell us it's a good one. You're hopeful. You have to be, but you never know."

And in fact, Jonagold, which he calls his first baby, has been a disappointment, at least in the United States. Even his own son, who runs the Way family orchard down in Pennsylvania, has not planted it commercially. Jonagolds are popular, but they haven't enjoyed the boom of new varieties like Braeburn,

Gala, and Fuji. Except in Belgium. He pulled out an article from a British fruit journal and read me the opening sentence: "If we didn't have Jonagold, I don't think there would be a Belgium apple industry." The quotation was from the director of Belgium's National Fruit Growing Research Station.

More than twenty thousand acres of Jonagolds have been planted in Belgium since the early 1980s—70 percent of all the apples grown there. Jonagold has turned the tiny country into one of Europe's leading apple producers. France, Europe's largest apple producer, has also planted Jonagolds widely. American apple growers have been far more cautious. "Apples are real slow to come along, and apple growers, of all people, are the most conservative lot you can ever find," Roger told me in lament.

Jonagold's phlegmatic fate may have a lot to do with the geography of apple growing. Washington State's Wenatchee and Yakima Valleys produce more than half of all the apples grown in the United States. The climate there is ideal for apples—bright, clear skies, low humidity, plentiful water from the Columbia River irrigation system—but not for the Jonagold. For some reason the large, yellowish-red apples sunburn in the high near-desert heat. Western Washington, where rainfall is heavy and the sky is cloudy, produces handsome Jonagolds, but because the growing region is small, most are sold to local markets. California's San Joaquin Valley, hotter than the Yakima, has even bigger sunburn problems. Michigan and New York have pushed other varieties ahead of Jonagold. Roger Way blames it on the commercial nurseries. "They control the publicity [for particular varieties]. We can't do anything about it."

To a point, apple breeding is still done the old-fashioned

way. Roger Way's successor, Susan Brown, has spent almost every Mother's Day for the last decade out in the Cornell orchards pinching blossoms and combing pollen. Unlike Roger Way, however, Susan Brown was trained in molecular biology as well as classical breeding, and she relies on the same techniques of genetic engineering that Norm Weeden and Stan Hokanson use to fingerprint the germ-plasm repository. But as a breeder—in fact the only full-time apple breeder in the United States—her objectives are radically different from theirs.

"There are over six thousand varieties of apple on record," she begins. "Why do we need a new variety? I always lead off my presentations with that question, because if I can't answer that question, I can't justify my position here. Our objective is to provide superior apple varieties." That key word, "superior," gets heavy stress. The questions, of course, are who decides what is superior, and what characteristics make one apple superior to another? Taste alone is not enough.

"You can have the best-tasting apple variety in the world, but if it is not making growers money, it is no good. We have apples that taste like apple pies. They're really spicy and wonderful, with complex flavors, but maybe they're not productive, or they bear biennially." She is clear about her commitments: only apples that are commercially viable can justify her time. In fact, Eastern orchards are littered with the remnants of once beloved apples that couldn't make any money for the growers. One of the most important apples of the early twentieth century, the Baldwin, has a rich, full flavor and is a reasonably good keeper. But because of biennial cropping, growers couldn't make reliable money on it. Our most popular apple in Kentucky, the Stayman Winesap, is a steady producer, but we lose 20 to 25 percent of the crop each year to cracking when late-season rainfall causes the flesh to grow faster than the skin; as we moved to quicker-producing dwarf trees, the problem grew worse.

Since her arrival a decade ago, Susan Brown has turned Cornell's breeding program upside down. Roger Way, Jim Cummins, and their generation of rootstock and apple breeders took known varieties that growers and apple eaters liked and then made tens of thousands of crosses to find improvements. Using that technique, hardly two centuries old, Cornell breeders produced Cortland, Macoun, Jonamac, Lodi, Fortune, Liberty, as well as Empire and Jonagold. Susan Brown, like other contemporary biotech breeders, looks instead for key genetic markers associated with the characteristics she wants new apples to have. Then she uses that knowledge to guide her breeding choices.

One of her first studies focused on the basic foundation of apple flavor: the ratio of sweetness to tartness, or, in apple talk, the acidity ratio. Traditional breeders would have done what Roger Way did with Jonagolds, which took more than two decades of testing. But by relying on genetic mapping and engineering, a breeder can make far fewer crosses, grow the seedlings for three or four years, harvest the fruit and measure its acidity ratio, and then discover the genetic codes that correlate with sweetness and tartness. All the rest of the seedlings that test out genetically as too sweet or too tart can be eliminated immediately from the test population. She can also use genotyping in the selection of progeny. Two parents that have only weak, recessive genes for tartness would likely lack character in a cross. Low sugar capacity in both parents would produce an apple that would be too sour. Those crosses could then be eliminated, saving time and money. Similarly, she can identify markers for strong or weak resistance to scab or fire blight within weeks of the seedlings' emergence; those that lack the genetic markers for resistance can be eliminated years before the disease would have shown up in traditional breeding programs. That, however, is the pedestrian side of apple engineering. More exciting to Susan Brown is the prospect of rebuilding the genetic foundation for already existing apples.

She takes me back to scab, the fungal disease that can defoliate an apple tree in hot, wet weather and cover the fruit with crusty cankers. Apple breeders have used the small flowering crab apple *Malus floribunda*, which is immune to scab, since the early 1900s. The primary scab-resistance gene from *Malus floribunda*, known as v_f, has been crossed into several apples, including Liberty, Goldrush, and Freedom, but it turns out that, while v_f is the most commonly found scab-resistance gene, it is not the only one. Other, rarer genes also provide scab resistance, and each one produces a different type of reaction, or signature, on an apple leaf when the fungal spores land on the leaf and begin to multiply. Some create a star, some a pinpoint, some a chlorotic wrinkle.

"The resistant gene is only incited when it is challenged by the fungus," Susan explained. To illustrate she pulled out a slide projector and flashed several pictures on the wall of her office. One showed a chocolate-colored fuzzy-looking blotch on a leaf, indicating that variety was susceptible to infection. "That is the fungus sporulating."

Next came a leaf with a pale, yellowish pinpoint: "That one has a gene for resistance."

Then another slide with a new leaf and a new pattern: "This is a different gene for resistance that was from material that came from Russia and causes a star-shaped lesion. These are different genes for resistance, and while many researchers are publishing papers about resistant versus susceptible [varieties], they're not looking at the reaction type. So what we've concentrated on is finding markers for all these different [resistance] genes." The search for those gene markers grew more urgent following an especially bad European scab outbreak in 1994, when varieties that carried the v_f gene became infected.

Apple breeders have relied on the v_f gene since it was identified in the *floribunda* crab in 1946. Ninety-nine percent of the world's scab-immune apples have it. And while the 1994 scab outbreak only hit the Europeans, it seems inevitable that that scab strain or one of four other "races" of scab that swirl in the air will eventually spread throughout the world. Quarantines and oceanic distance are simply no match for the modern transportation systems that have turned almost all local viral and fungal diseases into global diseases.

"The problem with the v_f gene," Herb Aldwinckle explained, "is that scab fungus can become virulent to it"—which is another way of saying that the fungus evolves its own resistance to the gene's attack upon it, just as bacteria have developed resistance to modern antibiotics. A rather dry, avuncular Brit who came to the Geneva station in 1970, Aldwinckle has worked closely with all the apple breeders' projects since then. He is, in effect, both the chief detective and the chief of preventive plant medicine. His pathology lab is where the genetic engineering takes place. Scab, powdery mildew, and fire blight are his target diseases.

Battling disease with a genetic arsenal is a game of numbers. Occasionally, a single gene controls the prospects for developing disease, whether the patient is a plant or a human being. Huntington's chorea, cystic fibrosis, Lou Gehrig's disease, and many cancers result from a misorganization or a malfunction within a single gene. In other cases, a particular gene will confer protection against disease, as seems to be the case with BRCA1 for women's breast cancer and a genetic anomaly that appears to prevent certain people from becoming infected with HIV, the virus that causes AIDS. Seldom, however, are we certain that having a gene—or lacking one—will guarantee protection. Viruses, fungi, and bacteria evolve continuously and rapidly, some growing weaker, some stronger or more virulent. It's that evolutionary virulence that has produced new forms of tuberculosis and gonorrhea, as well as the 1994 European scab outbreak that overcame the v_f scab protection.

Genetic-disease scientists like Herb Aldwinckle cannot guarantee certainties. Instead they calculate comparative probabilities. They ask: What are the odds that a particular pathogen can evolve the strength to overcome the effects of a particular gene or set of genes? How do those odds change if the plant or the person gets two, or three, or four genes that are resistant to the pathogen, or if, as in HIV or TB or scab, different strains of the pathogen exist? To fight scab, he wants to rely on redundancy, to bundle at least three distinct scab-resistant genes together and introduce them into the apple variety. "Theoretically," he said, "scab fungus could develop virulence to one of these other genes we're transferring into the apple. There's no reason why it couldn't. What we have to do, then, is transfer two or three [scab resistant] genes together, which makes it much more difficult for the fungus to develop virulence. If there's a chance of ten to the tenth [or one chance in ten billion] that the pathogen can develop virulence, then if you introduce two genes, it's going to be ten to the twentieth [or one chance in a pentillion—a billion billion] that the plant will become infected. With three genes, which is quite doable, it's one chance in a gazillion."

All geneticists are specialists in probability, but genetic engineers raise probability to poetic heights. Beneath those heights, however, they must wrestle with more mundane obstacles. Before Herb Aldwinckle gave me my crash course on apple engineering, I'd supposed that the tough problem was solved once he had located desirable genes in one variety so that they could be "injected" into another. In fact, no one has yet successfully transferred a scab-resistant gene directly from one apple variety into another. All of the so-called transgenic work has involved taking genes from other organisms and implanting them into the apple.

"The gene for scab resistance actually comes from a fungus that was developed to control root disease. The gene for fire blight [a bacterial disease] originally came from the gut of a Giant Silk Moth. Silkworms resist bacteria just as humans and plants do. Biochemists at Louisiana State found the protein and gene responsible, cloned it, and it's available to put into plants."

"Why would a Louisiana silk moth specialist be working on fire blight?" I asked.

"Well, they weren't. The protein, which is called Attacin E, is active against many bacteria. We developed it to control fire blight. In the plant the gene induces protection of the Attacin protein, which inhibits the bacteria. It has no mammalian toxicity. No plant toxicity. Makes the plant resistant to fire blight."

He handed me a color photograph of an apple limb whose leaves were bright and healthy.

"This is a Gala [a variety notoriously susceptible to fire blight] that contains a gene we inserted to make it more resistant to fire blight, and (*handing me a second photograph of a limb whose leaves were withered and black*) this is a regular Gala. It's blighted about to the [grafting] union." Then he showed me a second pair.

"This is a normal McIntosh that has been inoculated with apple scab. See the lesions here. This is a transgenic McIntosh, also inoculated, virtually disease-free."

One floor below and around a corner is the plant pathology lab where Aldwinckle and several of his colleagues conduct the genetic transfers. An array of glass-fronted refrigerators line the walls and contain hundreds of three- and four-inch culture vials. First he takes a tiny apple leaf from a sterile tissue culture and chops it into three or four small strips and places them on a moist, jellylike growing medium. Next he inoculates the leaf pieces with a soilborne organism called *Agrobacterium tumefaciens*, which can naturally implant the scab- or fire blight–resistance genes into the apple tissue. But the *Agrobacterium* doesn't only contain the anti–fire blight gene. It also carries a second marker gene that protects the tissue against an antibiotic called kanamycin. Kanamycin is mixed into the growing medium so that once the tissue begins to grow the ordinary apple tissue will die off, leaving the transgenic tissue healthy.

One of Aldwinckle's graduate students, a man named Joy

Bolar, pulled out a glass plate. "Here is a plate that has the normal leaves regenerating, and there is no kanamycin toxin in that."

"So you see the normal cells are not being killed in this case," Aldwinckle broke in.

Joy picked up another vial. "In this case I have the kanamycin [toxin], and it's not inoculated [with the protective genes]. So you see all the cells are dying." He reached deeper into one of the coolers, showing me a plate with a clump of tissue, some brown and dying, some with bright-green shoots.

"The tissue that hasn't been transformed is dying because of the kanamycin. And the little green shoots are coming from transformed cells that have received that resistance gene. They're growing on the medium, and they also contain the gene that we want to put in.

"Once the transformed tissue starts growing, we put it horizontal in the medium and it creates many clones." Later he will place the tiny clump of green into a new medium that will help it generate roots so it can be set in soil.

"So you see," Aldwinckle explained, "we can go from little leaf pieces to having plants in soil in seven months—which isn't bad, considering that when you do a conventional cross in May you won't actually get a plant like that for well over a year."

Creating these disease-free transgenic apples is quick, but it is not cheap. Geneva's plant pathology program has spent at least a decade developing the technique, learning how to control the tissue cultures and make micro-grafts with tiny, tender stems less than an inch long. Aldwinckle's colleague, Dr. Jay Norelli, made the first successful transgenic implant on the M.7

rootstock, a dwarfing stock that growers like but that has major fire blight problems. Work on the dwarfing M.9 has proved more difficult. Another pathologist, Jody Mills, is working on making a transgenic Gala, one of the most successful new apple varieties but one that is extremely susceptible to fire blight. New York apple growers have contributed more than a quarter million dollars, but once the basic laboratory and salary costs are added in, the bill for the project has likely exceeded a million dollars—and neither transgenic apple rootstocks nor transgenic apple varieties are yet on the market. None of them is likely to find its way into apple growers' orchards before the next century.

Some apple scientists liken these biotech changes to the Gutenberg revolution in printing. Aldwinckle reckons that it will take another four to five years before the so-called apple genome project has identified and mapped the genetic codes for all (or at least most) common apple varieties plus rare or still unnamed varieties from the Kazakh expeditions whose disease resistances are only now being discovered. Brown, Aldwinckle, and their co-workers might then fashion their own genuine "designer apples," by analyzing each apple tree's ancestry, the gene groups responsible for disease resistance, tree shape, cold-hardiness, sugar-acid ratio, firmness, shelf life, juiciness, vitamins, and possibly eventually the interaction of esters and phenyls that provide the subtleties of flavor. With luck those isolated genes could be introduced into varieties that apple eaters already know and like. A Golden Delicious could develop that stays as firm and crisp as a Fuji. A Jonathan or a Gala tree could be crafted to be free of fire blight. A dwarfing rootstock like MM.106 could be made that was less susceptible to crown rot (around the grafting splice) in wet soils. And expensive, toxic fungicides for scab could be eliminated. By unlocking and manipulating the vast complexity of *Malus*'s genetic heritage, modern science could take the final step in unraveling many of the most aggravating secrets harbored by the tree of knowledge.

Sometimes, however, biotech carries unintended consequences. During my visits to Geneva I had heard occasional reference to a terrible suicide committed by one of the research station's foreign collaborators, a Japanese scientist whose work had embarrassed the Japanese government and cost Japanese farmers an important market for their fruit. The man, Akio Tanii, held a position parallel to Herb Aldwinckle's at an agricultural research station near Hokkaido, the large timber and farming island in northern Japan. The Hokkaido farmers were well known for growing fine "nashi," or Asian pears, which they shipped to Australia.

Tanii had first gone to see the farmers in 1977, when they had complained of a new disease that was attacking their trees and turning the leaves black. Even then he suspected fire blight, which attacks pears even more vigorously than it does apples. After three years' research he published a scientific paper and concluded that the pears were infected with *Erwinia amylovora*, the bacterium that causes fire blight. Had Tanii reported the fire blight infection publicly in the popular media, it would have been devastating to the farmers, for Japan was one of the five big apple-growing countries in the world, including Australia, still believed free of the disease. The Australians were determined to keep themselves free of fire blight; they have banned fruit imports from any country known to have the disease—even though several scientific reports have shown that only the infected plant tissue, not the fruit, can spread the bacteria. Japanese agricultural officials solved the problem by convincing Tanii and his colleagues to call the disease "shoot blight of pear" instead of "fire blight." Then the Agriculture Ministry directed them to cease further research on the problem. All but two strains of the bacterium were sent to the Yokohama Plant Protection Station, where scientists were supposed to do follow-up research. Tanii's report was published —but only in a small journal that was never translated. There the story might have ended, but for a perspicacious scholar at Cornell named Steven Beer.

Beer, like Aldwinckle, is a plant pathologist and a world-renowned specialist on fire blight. "I was looking over an English translation of a new Japanese textbook on phyto-bacteriology," he told me, "and was curious about what it had to say about fire blight. Toward the end of the section the author went on for a paragraph or two describing 'bacterial shoot of pear—form species *pyri*.' So I wrote to him and said, 'Hey, this is very interesting. Can you tell me more about it and put me in touch with the guy who did the work?' "

That was 1992, and soon Beer and Tanii were corresponding. Tanii agreed to ship a sample of the bacteria to Cornell. By the following spring nothing had arrived, and Aldwinckle asked his son, Dave, who was living in Japan, to help out.

"There's a person down in an experiment station in Hakodate who's been a little slow in responding to our mail,"

Herb said. "Could you give him a call and remind him in Japanese to send us those fire blight samples?"

"I called and talked to Dr. Tanii briefly," Dave recounted in a long E-mail message describing the episode. "He seemed a bit harried (I thought it was because of this sudden call from a *gaijin* [foreigner] speaking Japanese; it happens), but said he would get sending. Mission accomplished.

"I thought no more of it for eighteen months, until I got an E-mail from Dad in late October 1995: 'Sad news. It seems Dr. Tanii has died, and it looks like a suicide.'

"I went bananas."

During the two years between Dave Aldwinckle's phone call and the suicide, Herb Aldwinckle and Steven Beer ran their own analyses of the *Erwinia amylovora* samples.

"We inoculated them onto pear, apple, and other fruits," Aldwinckle told me. Then he ran biochemical and DNA tests to determine the genetic nature of the bacteria. "We found that the bacterium was the same species that causes fire blight, but it definitely had some molecular differences from the *Erwinia amylovora* that causes fire blight. It caused disease on pear tissue that was similar to North American fire blight disease, sometimes even more severe, but on most apple tissue it didn't cause disease. On Jonathans, however, it caused disease almost as severe as our fire blight.

"The upshot was that we decided there definitely was fire blight in Japan, though the strains were different, and we presented this finding at a [1995] international meeting in Canada." Tanii's name was included in the presentation as a "cooperator." Australian quarantine officials attended the same meeting. They immediately quarantined the Japanese pears, which had been the only fruit imported from Japan into Australia.

Officials in the Japanese Ministry of Agriculture were especially upset and turned to Tanii's supervisors for an explanation. Tanii's colleagues told a *Los Angeles Times* reporter that the Agriculture Ministry hounded Tanii with relentless questioning and that outraged farmers would harass him with phone calls late at night. One group of farmers threatened to sue Tanii after the Agriculture Ministry ordered that all pear trees within forty yards of a fire blight infection be destroyed. A senior government official admitted to the *Times*: "I asked Tanii-san for an explanation. I told him he should get his boss's permission next time he does something like this." The whole episode was becoming a major embarrassment to Japan, and then it got worse.

Another American scientist, Rodney Roberts, was fascinated and puzzled by Beer's and Aldwinckle's findings. Roberts worked at the U.S. Department of Agriculture's Tree Fruit Research Laboratory in Wenatchee, Washington. Wenatchee apple growers, drowning in a sea of surplus Red and Golden Delicious, had just won entry into the Japanese market in 1994 after twenty years of pressure—but with a painful proviso. The American growers could only ship fruit from isolated "export zone" orchards that were surrounded by five-hundred-yard buffer zones that would be inspected by both USDA and Japanese agents. Once the apples were picked and boxed, they had to be held in separate cold-storage warehouses under sixty-day quarantine. Workers at the packing and storage facilities were required to wash their hands each time they entered or left. And the Washington growers would have to pay all the transportation and living expenses of the four Japanese inspectors for up to sixty-seven days a year—at a cost of $417,000.

The reason for all this precaution: fire blight. Despite clear evidence that the fruit could not carry the bacteria, Japanese officials set the growing and quarantine rules as a condition of entry. Once the Cornell fire blight report reached the Wenatchee Valley, the world's largest apple-growing region, the growers' reaction was predictable. Rodney Roberts, who had worked directly with the quarantine export program, called his contacts at the Japanese Ministry of Agriculture and flew to Hokkaido.

No, the Japanese assured him, they did not have fire blight. The Hokkaido pears suffered from a different disease, and in any case, they promised, the disease had been eradicated. Roberts walked through the affected orchards, where trees had been chopped down, but his official guide refused to let him take leaf and wood samples from trees that seemed to him to show ongoing fire blight symptoms. Instead, the government men took the samples and promised to run their own laboratory tests; they later reported that they had found no *Erwinia amylovora* bacteria.

Early on the morning of October 11, 1995, two months after Steven Beer made his report in Canada, Akio Tanii got in his car and drove away. He had told his boss, Fujio Kodama, the day before that the report had been submitted to *Plant Disease*, a major international scholarly journal, and that he feared relations with farmers would only grow more tense. He had also confessed to Kodama how worried he was about a meeting he had scheduled with a group of angry farmers. It was the farmers' anger more than government pressure that upset him so much that he offered to resign his job. He did not keep his meeting with the farmers. Later in the afternoon he staggered into the lab, his eyes dilated and watery. He had swallowed a bottle of pesticide. He died that night.

Fujio Kodama was distraught. "In the end I couldn't protect Tanii," he said.

"I cannot comment. He was not our employee," an official of the plant protection division of the Ministry of Agriculture blandly told the *Times* correspondent when she asked him about Tanii and his suicide. The ministry did, however, react to the affair. A week later it issued an order that requires scientists to get explicit permission directly from the Minister of Agriculture before removing a disease sample from its locale, thereby eliminating open collaboration with foreign scientists.

Back at Cornell, neither Steven Beer nor Herb Aldwinckle has had any further contact with the Japanese researchers. Beer asked Tanii's boss for the name of another scientist with whom he could continue the collaboration, but none was offered. Herb Aldwinckle was more blunt: "It seems to me they wanted to protect their industry, and if that meant suppressing this information, they'd be all for it. They'd suppressed just what this disease was. What we'd like to do is cooperate with them, sort things out, share our results." Then he added dryly, "I understand that some Japanese have replicated our research and the results were similar."

FIVE

Reds and Grannies
and Seek-No-Further

"Oh, it's a vicious business, apples," my host fumed. "I wouldn't ever grow apples again."

Now he was an innkeeper, puttering about a rambling, vaguely Greek Revival guest house alongside the Wenatchee River.

The city of Wenatchee, five miles downstream, is the world capital of apple growing. More apples were shipped from the Wenatchee rail terminal during the 1940s than from any other rail station in the world. Entire refrigerated apple trains headed eastward into the night for Minneapolis, Chicago, Detroit, Cincinnati, Philadelphia, and New York, and thence by ship across the Atlantic. Now they roll by truck. More than 115 million bushels of apples were picked and sold from the orchards of the Columbia River Basin in 1996. By the year 2000, the Washington Apple Commission expects to ship 133 million bushels.

Dennis, my innkeeper-host, grew only fifteen acres of apples, or about fifteen thousand bushels. He had already made his money in Seattle, in the office-cleaning business. He owned a Cadillac. His wife had grown up just north of Wenatchee, along the Columbia. They were timber people. Apples, he thought, would offer a quiet, pastoral retreat from city business. Crisp air. Crystal skies. He could tap as much water from the snow-fed Wenatchee as a grower could ever want. His trees were well-established producers of the most popular apples in America, Red and Golden Delicious. He didn't need to make a fortune, just a reasonable living.

I was sitting at the breakfast table on Easter morning, the only guest at the inn.

"See, I believe you ought to get paid for what you do."

His bitterness was as black as the impenetrable silence between his sentences.

"The only people that make any money in apples are the packers and the shippers."

Thirteen years he stayed in apples. Each year was worse than the one before.

"My best year I packed out four thousand boxes of X-fancy Reds. Twenty-five dollars a box. The packer wrote me a letter.

" 'I'm really sorry,' he says, 'but that thousand boxes we shipped to Chicago—couldn't get the money out of the buyer.' Something about how the [supermarket] chain went broke. Of course the packer wasn't out anything. I still had to pay five dollars a box for packing and fifty cents for shipping. That's over five thousand dollars I was out. He didn't lose a thing. I got nothing."

Those were his premium apples. The smaller apples that sold on supermarket tables at $1.69 for a three-pound bag brought him nothing once all the growing, picking, packing, and marketing charges were taken out. Columbia Basin apple orchards, he learned, were factories—"factories in the field" is how the late writer Carey McWilliams described the great California and Washington farms, vast planting and harvesting mechanisms that survive on tax advantages, mean margins, vertical integration, and global commodity orders. To succeed in these great apple factories a grower had to be a very sharp operator who was either big enough to run his own packing line or clever enough to stay just ahead of the fashions with new, hot varieties. Thirty years earlier, he could have ridden the crest of the Red Delicious boom that pushed nearly every other red apple to the edge of the table. But by the 1980s Red Delicious were in trouble, even though they accounted for more than half of the Washington State crop.

I had come to Wenatchee to meet the patriarch of the Washington apple industry. Grady Auvil had outlived and outperformed all his neighbors and competitors along the Columbia River. He had retired two of his children, and he remained president of the Auvil Fruit Company, which shipped its perfect fruit all over the United States, across the Pacific, and across the Atlantic. He was, on that Easter Day that I drove up to meet him, ninety-one years old, and, he told me with pride and a chuckle, he hadn't planted a single Red Delicious apple tree in forty years.

"1957. That was the last year." He paused for a moment to watch my reaction, then added, "The Red Delicious is dead, but it might take another twenty-five years to bury it." It was the sort of contrarian line Grady Auvil had been known for all his life. Other growers made fortunes, and often lost them, on Reds, and Reds remain Washington's leading apple. Half the orchards in the state are Red Delicious; in the words of another giant grower, it is the "flagship," the "golden goose" of the apple industry. Its characteristic long, toothlike profile finished by five prominent bumps at the blossom has been embedded into the world-famous Washington apple logo since the early 1960s. Its characteristic bland, chalky flavor has also been deeply embedded into public perception of the apple.

Not all Red Delicious are tasteless. The original apple, found growing on a farm in Peru, Iowa, from a seedling rootstock whose top had broken away, possessed a sweet, almost perfumy flavor. Vermilion stripes ran down over a creamy base. The farm owner, Jesse Hiatt, had tried to kill the tree several times, but each year the root sent up new shoots. At last he let it grow, and when it bore its first fruit in 1872, he became excited and named it "Hawkeye." Twenty-two years later, at the 1894 Stark Fruit Fair, C. M. Stark is said to have bitten into the apple and declared, "My, that's delicious—and that's the name for it." The tag identifying the apple and its owner had disappeared during shipping. Stark had to wait until the following year's fair, hoping that the same farmer would reenter it. Hiatt did, and the identification tag stuck to the shipping barrel. Stark bought the sole rights to propagate the "Hawkeye," renamed it "Delicious," and then spent three-quarters of a million dollars (at least ten million in 1995 dollars) promoting the Delicious to American growers. By World War II, Red Delicious had become the most popular apple in America. On our farm in Kentucky it found a strong following too, but those early Delicious bore little resemblance to the bloodred cartoon apples that covered the eastern slopes of the Cascade Mountains. Red Delicious are famous for producing "sports," or genetic variations, with dramatic shape and color differences. Soon nurseries everywhere were searching out the deep-red strains that orchardists would find on a single limb of a tree. Starking, Starkrimson, Red Chief were all dark, full-colored strains that stole the customers' eye away from the creamy stripes of the original Delicious.

More than any other apple, Red Delicious demonstrated the power of cosmetics in the fruit business. Even on our farm, customers will tell us how much they prefer "the old-fashioned Reds" to the fancy, perfectly shaped, perfectly colored ones at the supermarket. Yet without fail, whenever we bring in a bin or two filled with deeper, darker Delicious, those very same customers will go straight for the pretty apples and leave the pale striped ones behind. The farmers who moved out to Washington State around the turn of the century found out fast that the clear, dry air, cool nights, and long summer days would produce more beautiful apples than any other place in North America or Europe—and nothing was as beautiful as their ruby-like Red Delicious.

As late as 1960 the Washington Apple Commission still promoted what it called its "Big Six": Reds, Golden Delicious, Winesap, Rome Beauty, Jonathan, and Newtown. Except for Stayman Winesap, Cortland, and McIntosh, which were strong in the Northeast, Washington growers produced about the same array of apples as the rest of the country. The Jonathans filled the early market in September and October, followed by the Reds and the Goldens, while the Winesaps and Newtowns fought each other for the winter and spring months. Romes, the "Queen of the Baking Apples," lasted through autumn and winter. Growers could not afford to rely on a single variety if they wanted to sell their fruit all winter long, and that meant they needed hard, crisp keepers like Winesaps and Newtowns, or even earlier, Baldwins and Arkansas Blacks. Until around 1950 the medium-sized dark Winesaps, with their thick skin and dense, sweet-tart fruit, dominated the Washington orchards. Fifteen hundred miles from Chicago and as much as three thousand miles from the Atlantic coastal cities, the Washington growers had to have a late-keeping apple that stood up to cold storage without suffering "freezer blister" or internal breakdown. Winesap trains pulled out of the Wenatchee station as late as June and July of the year after they had been picked.

When I arrived in Wenatchee to see Grady Auvil, not a single supermarket carried Winesaps. At the biggest, fanciest produce section in town, Albertson's market, I asked the produce man, a young fellow in his twenties, if he had any Winesaps.

"We tried some of those a couple years ago," he said, scrunching up his face to remember.

"Not too big, deep red, tart and juicy but not too tart," I offered, trying to jog his memory.

"Yeah, I remember. Winesaps. We got a few in, but they didn't sell too good. I don't think we'll ever order them again."

When I drove up to talk with Grady Auvil, the fate of the Winesaps was still on my mind.

"Sure, I grew Winesaps," Auvil said. "Good producers. Good keepers. Let me tell you how quick they died."

Two things happened. The first was technological.

Botanists have long known that plants are continuously respiring, absorbing carbon dioxide and giving off oxygen. They have also long understood that if the rate of respiration could be slowed, or even halted, fruit would stop ripening. Refrigeration alone will only slow the rate of respiration. After four or five months, Red Delicious soften; Goldens lose their crispness after only two or three. To keep these softer varieties crisp longer, respiration has to be stopped, and the only way to do that successfully is to starve the fruit of oxygen and carbon dioxide. Farmers in the upper Midwest and New England understood as much a century ago, when they sealed their apples in barrels and submerged them beneath the ice all winter long. Come spring, they could hoist the apples up and find them just as crisp as when they were packed. Shortly after World War II, the English began to test so-called controlled-atmosphere refrigeration chambers, where oxygen would be displaced by nitrogen gas. Apples, they found, responded exceptionally well. Respiration ceased, the fruit stopped its release of the ethylene gas that promotes ripening, and the apples rested in a sort of suspended animation that left them almost indistinguishable from the day they had been picked. Grady Auvil, who had his own fruit-packing operation, was one of the first to see the new possibilities that controlled atmosphere, or CA, offered.

"At one time it was very difficult to keep Red Delicious beyond May or June in cold storage. Because you had to get them in condition that would keep, they had to be picked early, and they could scald [develop brown freezer burns]. Then they got so they could put them in CA, and they wouldn't get scald. So they could hold Delicious longer and longer and longer. So they grew more and more and more, and pretty soon they had Red Delicious to sell in August. When you had Delicious, they'd bring three times the price Winesap brought.

"So Winesap died real quick."

"But why?" I persisted. "When Winesaps are so much better, so much crisper and have so much flavor?"

He laughed with an old man's twinkle. "That's not what the public said. They said the hell with the Winesaps! The public likes a sweet apple, and Delicious is a sweet apple."

The proof was in the numbers. Winesaps fell from 50 percent of the Washington apple crop shipped in 1950 to 20 percent in 1965 to less than 5 percent in 1975 to less than 1 percent in 1985. At the same time Red Delicious rose from 50 percent of the boxes shipped in 1965 to 75 percent in 1987. Moreover, the total number of all Washington apples shipped during those two decades more than doubled, from 29 million boxes in 1965 to 68 million in 1987. Red Delicious ruled supreme.

Technology made it possible for Reds and Goldens to flourish, but a new style of targeted marketing for these remote high-desert orchards made it work. Washington growers had successfully pressed the state government to organize the Washington Apple Commission in 1937 at the depth of the Depression. Financed by a penny-a-box assessment on every box of apples shipped, the Commission grew into the most aggressive apple-marketing operation in the nation. Sales and promotion reps scoured the country, seeding newspaper food writers with feature stories about Washington apples, festooning produce departments with brightly colored promotional banners, and arranging cooperative advertising campaigns with supermarkets. New York, historically the nation's largest apple producer, had similar promotions, as did Michigan and Virginia. But the Washington growers faced a special problem. The very conditions that make the Columbia River Basin such an ideal apple-growing territory—rich volcanic soil, mountain air, dry desert slopes—have also separated it from large population centers. While around a third of Eastern apples are sold directly or in roadside stands to consumers, Washington's growers have almost no direct sales and count exclusively on national wholesale brokerage. Almost from the beginning, successful growers had to limit the number of apple varieties they grew. Eastern orchards often sold fifteen or twenty different kinds of apples, but wholesale buyers seldom wanted more than a handful or even two or three varieties.

As promoters, the Washington Apple Commission had to contend with several dozen different brand labels, almost all selling the same "Big Six" varieties. A solution came in 1961 from Container Corporation of America, the box manufacturer, whose designers produced a Commission trademark for all Washington apples: a single, jaunty apple, in either red or yellow, drawn in the unmistakable shape of a Red Delicious. "Watch for it," a promotions writer for the *Wenatchee Daily World* advised in the paper's annual spring apple supplement. "You'll see, over the course of years, the rise of a major trademark in the American food business."

Every box, every newspaper ad, every supermarket banner and decal, and eventually every individually packed apple shipped from Washington State—whether it was a Red Delicious or a Winesap, a Rome Beauty, or later a Fuji or a Gala

—was stamped with the Red Delicious emblem. The choice of the Red Delicious as the state emblem was far from accidental, Dick Bartram told me. Bartram started out in the 1940s as a horticultural extension agent and went on to become a national marketing specialist, visiting major stores all over America.

"No, it wasn't accidental at all," Bartram said. "The emphasis was placed on Red Delicious due to the fact that it was rapidly increasing in value, and the apples had to be sold. The Commission was really concerned because the growers kept on increasing the number of acres going to this variety. The Reds and the Goldens just pushed everything else out of the market." Superficially, at least, Bartram is correct. Red Delicious made a flashier-looking apple than Winesaps, and their mild, bland sweetness proved thoroughly inoffensive. As Cornell fruit breeder Susan Brown expressed it, Red Delicious came into flower when the prime breeding and marketing objective was to find "inoffensive tastes" in food, tastes that would alienate no one and presumably claim the broadest possible market share. It was the era of Wonder bread, Cool Whip, Big Macs, and Campbell's condensed soup casseroles: food marketers were learning the lessons of brand loyalty. The Washington apple—either Red or Yellow—was easier to market than a matrix of subtle tastes with old-fashioned names that no longer resonated with the children of the sprawling American suburbs. Again, even in our remote Kentucky hill country, young adults of the 1970s and 1980s would walk into our sales shed with little or no knowledge of particular apple varieties. They wanted sweet or tart, hard or soft, red or yellow, and they counted on us to tell them what they should buy. They no longer knew particular apple flavors. They no longer knew which apples would last the winter. They had no idea which were better for sauce, which were better for baking, which were better for stewing or making pies.

Trademark imaging proved a brilliant stroke for the Washington growers on another front as well. Heady competition among the national and regional supermarket chains in the 1970s utterly transformed American produce marketing. As Kroger and Safeway rapidly outpaced older, more sedentary supermarkets like A&P, they began negotiating contracts for hundreds, sometimes thousands of trucks and boxcars of a particular item. To cut those deals, the packers had to be able to offer consistency on a mass scale. They installed packing lines that carefully weighed and measured each apple passing over the line while "computer eyes" scanned the fruit to be sure it met color requirements. Goldens with russet were dumped into a second grade. Reds that registered too pink fell out. Brokers simply couldn't afford much variation, and, because of Washington's climatic conditions, almost no other growers could match the Yakima and Wenatchee Valleys for consistent cosmetic elegance. Not until the 1980s would computer software become sophisticated enough to enable the chains and the shippers to track large, complex mixed orders easily. The marketing system that had helped build the Red Delicious into the national apple could only handle mass volumes by relying on mass uniformity. Red Delicious had simply been the right apple at the right moment in popular taste at the right stage in the development of mass marketing. Or, as Dick Bartram told me, Reds "crowded Winesaps out of the market" as the market itself changed. By the late 1980s, however, a new market shift was emerging. Red Delicious's share dropped from 75 percent to just over 60 percent in 1995; insiders predicted it would dip to 50 percent early next century. "Now," Bartram said, "we're seeing the same thing—except that Red Delicious is the victim."

Grady Auvil, the nonagenarian Puck, has been one of its chief saboteurs. His weapon was a sour green apple called Granny Smith.

"I helped push Granny Smith," he said modestly, waiting to see if I'd done my homework.

"You *were* Mr. Granny Smith," I proposed.

Another tinkling laugh broke out.

"I researched it. I shot my mouth off, and now they're growing it. It'll be a good apple fifty years from now."

Auvil used the word "research" advisedly.

"I looked at Granny Smith for about twenty years before we started growing them, since they first started importing Grannies in the late 1950s. There was no place you could buy the trees. Well, I had a friend in Canada who had wood from New Zealand. I grafted a few trees. It takes time to get an apple into bearing. We looked at the apple. We stored it, and we ate it. It wasn't till up in the 1970s that we went out to promote and propagate it." He harvested his first crop in 1975.

"We were the only guys around here with them. We had a free Granny market for years. We'd get five, six dollars a box above the Red Delicious price. We did all right."

Very all right. Red Delicious were bringing sixteen to twenty dollars a box for most growers in the 1980s; Auvil was collecting 30 to 40 percent more for his Grannies and shipping five to six hundred thousand boxes a year. The Grannies enabled Auvil Fruit—trademarked "Gee Whiz" brand—to hire its own national sales force. By the late 1980s along came Galas, which took them even further away from Reds. Well into his eighties, Grady Auvil was still working on his dream apple.

"I'd looked at Fujis for ten years before we got into them. I got some Japanese [nursery] catalogues and had them translated into English so I could read them." He paused for a moment, instructively: "Whenever you listen to a nurseryman, you open yourself to big errors. They only show you the best side of what they're selling. They talk out of the wrong side of their mouths." Having made that clear, he resumed the Fuji story.

"I got specimens of all those [Japanese] apples and grew them. Some were pretty good. Most of them weren't."

The first Fujis went into the ground in winter 1973–74, on M.26 dwarfing rootstock. Two years later he had a crop. By 1990 Fuji had become a Gee Whiz gold mine, packing out at fifteen hundred to two thousand boxes per acre, averaging more than forty dollars a box.

"This is the whole ball game in growing fruit now: high tonnage and good quality. If you get both of these, you're in business. You look at Grannies. The average [in 1996] is eighteen dollars a box. We probably average twenty to twenty-two dollars, and we've gone as high as thirty. We've gone as high as sixty-five for Fujis.

"There's one rule in this free economy. The people who are buying are always crowding the price down and tend to degrade the whole situation. The farmer, being at the bottom of the barrel, is the guy who always carries the load. The only way you can avoid that is to have something that's in heavier demand than supply. That is where we work all the time, so we can jack the price up."

High-price apples like Grannies and Fujis carry a catch, however: they are expensive to grow. Fujis are especially difficult. Until recently, most strains would not develop pretty red cheeks without cool weather, which has posed problems for California and Midwestern growers. The trees themselves tend to be unruly in their shape. Humid weather or spraying under the wrong conditions can ruin their finish. Most troublesome, the apples ripen unevenly over a three-month period from late August to November. Pickers often have to go back to the orchard three or four times to select for color and ripeness, sending labor costs into the heavens. And pickers must carry small clippers to trim each apple's long, sharp stem to prevent it from puncturing and ruining the fruit that rests next to it in the picking bins and packing boxes—all of which means it takes at least triple the number of hours to pick a Fuji orchard as a Red Delicious orchard. Or, as Auvil Fruit Company's orchard manager told me, "You can grow a box of Delicious for half the cost it takes to grow a box of Fujis."

Cost, however, falls low on Grady Auvil's list of concerns. "You look at two different things in life. I never looked at ways to save money. I always looked at ways to make money. Same with wages. We [he and his brothers] started out working for other people in the Depression, so I always liked to pay good wages. Our pickers make a lot of money. We pay thirty to thirty-five dollars a bin [twenty-five bushels] for Fujis, and twenty-five dollars for Grannies. That's five to ten dollars above standard. A lot of the other growers think we're crazy.

They said I'd be out of business years ago. Our pickers do not deal in bruised fruit, and we get two to three dollars above the market per box, sometimes even five to six."

I mentioned a major California Fuji grower I'd visited who only paid his pickers minimum wage—at that time $4.75 an hour—about half the rate Auvil Fruit was paying to harvest its Fujis. He shrugged. "I would not like to be in business if I had to fight nickels for wages. I suppose that's why we're always doing new things. The most important thing for us to do is not to worry about saving money. The most important thing is to analyze what's going to be here ten, twenty years from now."

"So what are you planting now?" I asked.

"Fujis! Fujis! Fujis!"

"And what about the next century? What will the hot apples be in 2012?"

"Granny Smith and Fuji. Fujis and Grannies will last at least until 2050."

Twice each year, down in a valley along Virginia Route 20, halfway between the two great monuments created by Thomas Jefferson, Monticello and the University of Virginia, Tom Burford holds an apple tasting. Dozens of varieties of apples fill wicker baskets that have been lined up on long banquet tables. Some of the varieties grew in Mr. Jefferson's original orchard, which has been restored with Tom Burford's help. Most come from Burford's own nursery, an hour away in Monroe.

"There are collectors who want to have as many different

kinds of apples as they can get," Burford begins. "I want as many different flavors as you can get."

Tom Burford—the r's in his voice sound as soft as a Tidewater breeze at the end of August—grew up in a Virginia where rhetoric, elocution, and home fruit grafting were as common as Nintendo is today. The Burfords, he reminds his listeners, were growing apples as long ago as Mr. Jefferson was, and given the travails of Monticello, one presumes with considerably more business acumen. He lends his pocketknife to an assistant, who begins to slice the first apples to be tasted.

Before passing the slices around, Burford instructs his audience of about fifty Monticello visitors—who have made their reservations weeks in advance—"how to taste an apple." The apples will be placed on paper plates along with small chunks of bread to clear the palate. Each person has been given a sheet of paper with each apple's name on it and space to write in a numerical ranking. Peter Hatch, the director of the gardens at Monticello, has already told the gathered tasters how once there were great competitions among apple growers, how each county had its own prized varieties, and how at fairs the apples "were rated like great restaurants or rock and roll singers."

"Taste," Burford continues, "is subjective." He advises them to close their eyes and allow the apple juices to wash the mouth, because "there is no need to look at your neighbor's sheet and cheat."

"Let's start with one of the oldest known apples in the world, the unsung White Winter Pearmain," or, in the music of his drawl, the Weyete Wintah Peaehmen. The White Winter Pearmain, sometimes known only as Winter Pearmain, came to America from England and may have been one of the apples the Romans planted on the hillsides above the Mediterranean. The apple—sweet, almost honey-like, with a clean, lightly acidic residual—was a favorite of nineteenth-century growers. Its blossoms are lightly bluish, which has also made it a prized ornamental. "It would keep late into the winter, but it was downgraded commercially for its rough appearance.

"We can jump from early England to early America, where one of the many seedlings developed by the colonists was the Golden Russet. Farmers used it for juice to make [fermented] cider and brandy. Economically, it was very profitable. The Golden Russet appeared as an orchard seedling in the early 1700s and is considered one of the classic American varieties. It's good for eating out of the hand, but it's excellent for cooking, and it's one of the few apples good enough for making cider alone."

Picking one up and rubbing his thumb lightly across its skin, he added, "Notice its sandpapery feel. I've seen people pick them up and drop them saying, 'Oh, that's rough!' " He pauses for effect: "I predict they'll be the next big apple."

Next he chose a favorite from the 1920s called Mother, which "my great-grandfather Burford" cultivated commercially because its late bloom left it safe against the frost. Mother is a large apple that ripens at the end of summer and has stolen the praise of many an apple connoisseur. The British pomologist Edward Bunyard, who included it among his "top ten," wrote that Mother bore the "flavour of pear drops," while the nineteenth-century English fruit encyclopedist Dr. Robert Hogg detected a "balsamic aroma" in it. Others found a "hint of vanilla" in the Mother. But its soft yellow flesh spelled its doom, for it could be neither stored nor shipped.

For his third selection, Burford leaps forward to Grady Auvil's passion, the gold of the 1990s, the Fuji, which, as it happens, owes its own debt to Mr. Jefferson. Fuji was one of the
last Japanese crosses to be released without a Japanese government royalty. It developed from a project begun in the 1920s,
when Japanese horticulturists came to the United States and
took cuttings from many of the most popular American varieties. Fuji appeared in Japan in 1939. Its parentage, however,
was Ralls Janet × Red Delicious. The history of the Ralls is
fuzzy. It is credited to Caleb Ralls of Amherst County,
Virginia—"the Rallses were neighbors of mine," Burford
allows—but it was also thought to have been brought from
France by a M. Genet, a Protestant minister, as a gift to Jefferson during his Presidency; it is sometimes known as the
Geniton or Jefferson Pippin.

Pedigree is incidental to Tom Burford's remarks about the
Fuji. He is a Fuji enthusiast, but he is pointedly not enthusiastic
about what large commercial growers and nurserymen are doing to Fujis. "One nursery even has a reward out: 'Find me
the reddest one you can find!' they say. That's what has happened to many good apples, like the Red Delicious, originally
called Hawkeye."

(The pursuit of red pigment, he is convinced, nearly always
comes at the cost of taste. When nurseries seek out deep-red
sports, as they have with Fuji and Jonagold, and famously with
Red Delicious before them, taste does seem to be sacrificed to
cosmetic flash—or so a study published in the American Pomological Society quarterly, *Fruit Varieties Journal*, reported in
1989. A taste panel of 271 consumers at Penn State University
sampled nine strains of Red Delicious, among them: Starking,

Starkrimson, and Red Chief. The deeper and more perfectly red the apple, the more bland and insipid its flavor. I mentioned the color-versus-taste equation to Grady Auvil. He did not reject the proposition but instead made a note to undertake his own tests, both in the laboratory and with individual tasters. His supposition, however, was that growers use early coloring to pick their apples before they are ready so that they will store better under refrigeration. With both Red and Golden Delicious, premature harvest has devastating effects on flavor. Should there turn out to be an inverse relation between genetic variations that produce intense color and intense flavor, the genetic engineers would find themselves with an especially interesting problem, for both apple flavor and apple color appear to be controlled by a complex array of genes.)

"One of the 'olive' apples—you either love 'em or hate 'em—is the Westfield Seek-No-Further," he says, picking up a medium-sized, oddly shaped apple with red stripes. Found sometime in the mid-1700s in Westfield, Massachusetts, near Springfield, the nutty-flavored apple convinced its enthusiasts it was the *Malus sumum* of New World apples, and for much of the nineteenth century it claimed a strong following as a New England home-orchard apple: "They'd bite into one, say 'Pretty good!' then either smile or spit when they'd get that little astringent aftertaste."

As the plates of apple chunks pass down the aisles like Communion trays, Burford supplies a flood of anecdote, social history, and sharp pomological commentary. The arrival of the railroads in the 1850s enabled prosperous farmers to pack and ship carloads full of barreled apples up East, and as a result, Winesaps, originally a New Jersey cider apple, blanketed the

Shenandoah Valley, driving out the old, reliable Virginia Beauty. The Hunt Russet, dating back to the 1680s, was such a good keeper it would last for a year in a common root cellar. Baldwin, the great but difficult (it gives a full crop every second or third year) apple of New England, was wiped out in the winter of 1934, when the temperature in late February dropped to negative forty degrees following an early warm spell and many old trees burst open from the frozen sap. Within a year or two, New England farmers were ripping out their ghostly dead Baldwin orchards and replacing them with a smaller, sweet red apple from Canada called McIntosh.

Forty-four different apples have been sliced and circulated by the time Burford reaches his final four. "We will conclude by tasting four apples from Mr. Jefferson's orchard," he announces. "The apple we'll be tasting first is Pomme Gris, the gray apple. It is a russet. There are a couple of strains, but this is probably the most authentic one. You'll notice it has a very intense flavor." He picks up a second, rough motley fruit. "The Roxbury Russet is very similar to the Golden Russet (the second apple of the tasting), but it is much flatter, it bears heavier, ripens earlier, and also comes from New England, though about a hundred years earlier, probably at the beginning of the seventeenth century. It was very popular in New York State because it was such a fine keeper." Again, another dramatic pause as he holds up a greenish-yellow apple.

"Newtown Pippin. Newtown, Long Island. The Princess Nursery. That's where Jefferson got his apple trees. The English in the time of Queen Victoria became *addicted* to them. There's no doubt that the shipping time contributed to their flavor [even in the 1960s, when it was still a major apple in the Pacific Northwest, the Washington Apple Commission promoted Newtowns as the late-winter and spring apple]. They could command three times the price of any other apple." Newtown became the first genuine American export apple, leading growers to plant it throughout Albemarle and Nelson Counties, along Virginia's James River. Rankled by the high commissions New York brokers were charging, the Virginia growers switched to the port of Baltimore and renamed their fruit Albemarle Pippin to compete with the New York shippers. Mineral depletion in the soil, however, proved a more intractable problem than brokerage fees. These great, gnarly old trees that could reach a height of forty feet and bear for seventy-five or eighty years were steadily leaching all the calcium from the soil; as they did so, each season's apples became marginally softer and began to lose their fine keeping qualities. Twentieth-century orchardists add calcium to their spray mixes, but earlier farmers had neither the foliar- and soil-chemistry assays to identify the problem nor the technology to correct it. Gradually the Newtown slid toward oblivion.

At last, the final selection.

"It is said repeatedly that Jefferson preferred the Esopus Spitzenburg to all other apples. It's long been one of the classic American varieties [found in the late 1700s at Esopus in Ulster County, New York]. Jefferson struggled to grow it here in Virginia. It makes a very friendly tree. The limbs come out from the trunk at just the right angles. But the problem was that it's terribly subject to fire blight, and in this hot Virginia climate, well . . ."

Tom Burford's caustic broadsides against the "monoculturists" and the "cosmetic" forces in contemporary horticulture, seasoned by his genteel, Old Virginia style, have made him the prince among antiquarian apple enthusiasts. He is, however, intensely practical. He dismisses out of hand romantic notions that apples can be grown "organically" in the hot, humid Eastern climate, but he is always willing to talk. To a degree he sees himself as a quiet agricultural diplomat between the conventional and the organic growers. "Commercial growers come to me very often, and they say, 'Tell me what is this about Integrated Pest Management [a technique for reducing chemical applications by careful monitoring of orchard pests and greater reliance on natural predators].' They are asking for the wrong reason, but it is well they do; they are doing it because their chemical bills are escalating so fast they need to do something about them. On the other hand, some of the die-hard organic growers come to me and say, 'Do you think that if I use Imidan [a key general pesticide] I could get a better yield, because I'm getting probably 10 percent salable fruit?' And I say, 'Sure, try and see if you can't get 20 percent,' and when they get 40 percent, they are pleased!"

Even more noisome than organic ideologues are the obsessive antiquarians.

"I debunk the notion that antique varieties are the solution to the apple industry," Burford declared one day while I was visiting his nursery at Monroe. "They are apple addicts, people who want to experience all these varieties and usually with an undercurrent of (*sotto voce*) 'Now, how can I be an entrepreneur and capitalize on this.' It's this notion that all heirloom varieties are superior, it's the Washington/Jefferson idea of these great, pure, perfect models. And the fact is 90 percent of the heirlooms could be called 'spitters'—one bite and you spit."

For Burford, the heirloom fetishists and the giant commercial monoculturists have joined in an unknowing collusion that has worked to impoverish contemporary apple growing. "Look at all the agricultural universities," he went on. "What kind of plant-breeding programs have they got? The apple addicts who keep pushing antique varieties have really been a disincentive for plant breeding, for deliberately making new varieties, using the new [genetic] technology." The fact that Cornell alone has maintained a full-time apple-breeding program while there were nearly forty-five such programs half a century ago is ample evidence of the problem. At the same time, the emergence of a global trade in apples, as in other commodities, has skewed the search for new varieties to a few very limited criteria: hard, pretty fruit that both last well in storage and have enough shelf life to hold their crispness for a full week at normal room temperature. Perhaps the prime example of the new global apple, he agreed, is the Pink Lady.

I'd tasted a Pink Lady at the Sierra Hills Orchards outside Stockton, California, in early October 1996. Mark Lewis and his family were the first to strike gold with Fujis, in the early 1980s, when they secured major contracts to supply the Taiwan and Singapore markets. The day I arrived, Lewis had just dispatched the final truckload of fancy Fujis to the port of Oakland for a transpacific shipment. Sierra Hills ships more Fujis than any other single grower in the United States—70 percent of them to Asia. But his market window is small. His Fujis ripen three weeks ahead of the Wenatchee Valley Fujis that Grady Auvil grows, but they never develop the rich, red blush of the Washington fruit. Soon, Lewis acknowledged, the Washington growers will get an early Fuji that is prettier than his, and then, he said, he would be squeezed out of the global

Fuji trade. In its place, he has pinned his hopes on Pink Lady. It ripens early, withstands the heat of the San Joaquin Valley, stays hard, and is full of juice.

I told Tom Burford about the Pink Lady that Lewis had offered me—how it met all the major criteria except one. I thought it had very little flavor (although I'd heard that Pink Ladies grown in the East were much better).

"Pink Lady is being planted like no other apple variety in New Zealand and Australia," Burford said, adding that he kept regular correspondence with the apple men Down Under. "None of the growers there that I have been in contact with really say, 'Well, here is a wonderful-tasting apple!' This is really an apple they're growing for the export market."

Bland Pink Ladies. Overplanted Red Delicious. Over-colored Fujis. All of these commodity-driven phenomena destined for the "export trade" have, in Burford's view, im-poverished the entire American food supply. They have pro-duced ordinary, uninteresting apples that simply can't compete with the high-salt, high-fat, high-sugar, high-packaged items that have saturated the snack-food market. "Americans are starving because they don't produce the proper food," he railed.

Starving?

"They are not eating properly. In a few decades, there are going to be government commissions to study why nutrition in America is so poor."

I wasn't sure that "starving" was the right word, but Bur-ford's choler did recall the reaction I'd had a couple of months earlier at the annual Kentucky State Fair. We had taken our own fresh cider to the fair, run it through a rotary freezer, and sold "cider slushies" by the glass. Customers loved them and came back for seconds, but it was clear that even at a huge agricultural fair—some three-quarters of a million people bought tickets—fewer than a quarter of our customers had ever tasted genuine, fresh farm cider. Worse still, easily two-thirds of the families, including the children, who passed by the farm market tent were—no other word—obese. They were packing their gullets with deep-fried "onion blossoms" and funnel cakes and swilling down thirty-two-ounce mega-cups of Coke.

Apple tasting at the fair had disappeared so long ago that no one had any memory of it. When I was a boy we and another dozen orchards had shown our apples at the fair. One of my fondest childhood memories was scouting through the orchard in early September for the most beautiful early Goldens, Reds, Grimes, King Davids, Jonathans. It was too early to find a pretty Stayman or a Rome with any color, but we won our share of blue ribbons. We never questioned the fact that already the judges' only consideration was cosmetic: how well colored the apples were, how smooth their finish, how true to type their form. No one ever tasted the apples. They were too immature to have any taste. Sometime in the late 1960s the fair was moved back to the middle of August, so only a few growers down on the Tennessee border had anything ripe enough to show. The sort of genuine, heated competitions among local farmers for the best pie apple or the juiciest eating apple was not only lost: it had become inconceivable.

The giants of the U.S. apple industry know they have a problem. Each year the hundred or so most powerful apple men (and they do seem almost all to be men) gather at the winter meeting of the U.S. Apple Association to track industry prospects. At the 1996 meeting in New Orleans, everyone was worried about apples' shrinking share of the food market. Desmond O'Rourke, an agricultural marketing specialist at Washington State University, ran down the numbers. Apple consumption had been "flat for the last thirteen years, and real value at the grower level has stayed below $1 billion since 1980. We are not capturing a bigger share of the consumer dollar."

Americans consume a little more than nineteen pounds of fresh apples a year—far less than any European country. The French eat thirty-three pounds, the Germans almost forty pounds, the Italians slightly more than fifty-seven pounds, and the Benelux countries devour nearly sixty-four pounds of apples a year. Yet overall, Europeans, who are both the world's second-biggest apple growers (after China) and easily the world's largest apple importers, are steadily cutting back their appetite for apples—down from forty-six pounds per person in 1981 to forty-two pounds in 1993. Meanwhile, the world's total apple supply continues to rise, from 38 million tons in 1980 to 48 million tons in 1993 to a projected 58.3 million tons by the year 2000. The developing countries of the world—Turkey, Russia, Peru, and Brazil, plus several Asian nations—have taken up the slack, but most consumers there cannot afford fancy apples; what they want is what they can afford: cheaper, smaller apples. If the orchardists of the Northern Hemisphere are going to find new customers, they will have to offer apples that compete with other fruits—especially citrus, bananas, and other tropical fruits—but, even more important, with cheap snack foods.

"Potato chips," O'Rourke told the apple men, "are falling

in price more rapidly than Red Delicious. This really is one of the key competitors to apples. You have to remember that last year [1995] 47.2 percent of every food dollar was spent on food eaten away from home. The real competition will not be so much between different kinds of fruits or different states but between the world supply of snack foods and the world supply of fruit. That is where the apple industry is losing the battle. Doctors and nutritionists continually advise people to eat more apples. If you look to the developing countries, their consumption will rise as their income rises, *but*," he warned, "the people we're really up against are Frito-Lay, Coke, and Pepsi, the mega-marketers of snacks."

On that point Grady Auvil and Tom Burford couldn't agree more. Both men, widely sought out for advice and prognostication on the future of the apple, are revolted by what is offered for sale in today's supermarkets. "If you bought apples from the stores," Auvil told me, "you'd find that at least half the time you were getting an inferior product"—by which he meant the small, dry, cardboardy Red Delicious his neighbors pump out. He was, on the other hand, optimistic that newer varieties have broken the Delicious's grip on the market and that demand for good apples will start growing again. "If we were to supply the American people the kind of product they appreciate, even 90 percent of the time, we couldn't grow enough apples in this country."

Burford, who goes by the sobriquet Professor Apple among his admirers, is far less optimistic. He acknowledges that men like Auvil have helped bring first-rate apples back to the American market, but he believes the whole global commodity system is everywhere eroding food quality, with the result that the customer has been caught in a squeeze between the faddish apple addicts and the brokers who trade in gross container shipments. "My agricultural philosophy embraces the fragmentation of the food industry so that once again we get into back-yard food production, local market competition, and the reverence for the commodities. There is no reverence for apples today.

"If a local guy says, 'Hey, I've got forty bushels of King Davids, I'm going to take them to the market and sell them,' he couldn't. Nobody would buy them. Not that they are not good apples, but because he doesn't know how to sell in the modern marketing world. He doesn't know how to say, 'We'll do a little tasting,' and pass a tray around. He sits and waits for the consumer. Well, the consumer doesn't know anymore. Look at how minuscule apple consumption has become in America. It has to be done at the grassroots. Consumer education. The organizations like the U.S. Apple Association, they do not understand. These are highly bureaucratic organizations that reach out and touch the wrong people. They don't know what is really happening, and they don't care so long as they justify their jobs.

"More and more people have to come to apple tastings. They have to discover apples, and then go to their markets and demand them."

SIX

Cider: Lifeblood of the Heathen Apple

Étienne Dupont's manor house is not grand, but it is certainly elegant. A driveway of fine, crushed tan gravel leads straight down a long green, past a weeping willow, and around the boxwood to a double oak door. A drainage slough and a young orchard lie to the left. A long brick barn, once a stable, its white doors surrounded by grand stone arches shaped like horseshoes, stands to the right of the green. The house itself, constructed of eighteenth-century painted brick capped by a blue tile roof, sits, like many Norman houses, a stalwart anchor to the estate.

I was late. I had called M. Dupont a day or two earlier from Paris and explained my interest in traditional French cider making. The drizzle and fog of early November had made the drive longer than anticipated. Mme. Dupont, a trim, sharp-featured, handsome woman of about seventy, invited me into her living room, where she had been working at stitchery beside the fire. Her son, she assured me, would return soon. One can easily fall into tourist clichés in Normandy.

"We were to meet at nine-thirty," Dupont began when he walked in the door, hurried but cordial. Country rustic is not his manner.

"Well, so . . . Just now it is not possible. You are here how long? Just the day?" Sizing me up quickly, he decided we would spend our time much more efficiently conversing in English. He had, after all, spent the better part of a decade working in London as a senior officer in one of the great French banking houses.

Étienne Dupont, I had been told, had a mixed profile. Like most Norman families, the Duponts had been cider makers for generations. They had operated a conventional Norman farm, raised cattle, kept horses, grown crops and forage. Though not possessed of enormous wealth, they were solid pillars of the educated, literate agricultural bourgeoisie. When Étienne returned from England in the late 1970s, he found a farm and an industry in gradual but unmistakable decline. Cider consumption had fallen steadily since the end of the war as the region's population fell. No one in Paris or Lyons, the great culinary cities, considered cider anything more than a rustic relic, and in an era when the passion to be *moderne* had overtaken all else, a quaint, fizzy apple wine carried little cachet. Calvados, the distilled brandy made of cider, still lured sophisticated palates, but it was no competitor for cognac or its trendy cousin, Armagnac. If the still young Dupont wanted to salvage the family resources through cider, he faced a very large challenge—and not one that many of his banking colleagues thought likely to succeed.

"I am the last of the Mohicans," he told me when we met later that day. By which he meant two things: he saw himself as one of the last practitioners of classic cider making, and he saw himself as leading a life-and-death campaign to save the culture of cider making that the Duponts and a few dozen small artisans practiced in the Pays d'Auge district of central Normandy. That was in 1992, and the prospects for cider were hardly promising.

A fierce winter storm had uprooted half the French cider trees in 1987, and neither the farmers nor the Ministry of Agriculture had made much of an effort to replace them. The same storm had ruined cider orchards all over Somerset and Devonshire in England and left growers similarly resigned. In America, cider, the most popular beverage at the birth of the Republic, had disappeared but for a very few local producers in New England and the Northwest. Most Americans supposed the word referred to just the raw, sweet unfiltered apple juice sold at farm stands. Never had a fermented beverage fallen so far, so fast, from popular favor.

I first tasted genuine cider at a Breton café off the boulevard Montparnasse in the early 1970s. The waitress brought a champagne bottle and poured a clear, amber-gold stream into an earthenware drinking bowl. It erupted in a cloud of crystal bubbles. She plonked the heavy bottle down and walked off. I shifted my eyes between my dining mate and the bottle.

Were we to drink the whole thing and still walk soberly off to the movies? My friend knew no more of *cidre* than I did. I raised the still lightly fizzing bowl to my lips. The bouquet of apples was unmistakable, though not of any apples I had ever tasted at home on the Kentucky farm.

This bouquet was sweet like the vaguely rosy fragrance of apple blossoms, but just as present was a darker, richer, moody smell of autumn. Its aroma bore neither the exuberance of a grand, glittering champagne nor the sadness of a fine old sherry. It tasted instead, if I can say it, of an early melancholy. At first, I confess, I did not like it very much. I had supposed, like most first-time cider sippers, that cider would be simply another sort of wine, which it is not. Like your first oysters or artichokes, cider requires its own neural pathway into gastronomic memory. It must be taken altogether on its own terms, its own emotions, its own colors, its own strengths and liabilities. (A cider hangover headache, I have come to learn as well, always hits me in a spot of the brain that is affected by no other beverage, fermented or distilled.)

An expertly made cider fills the mouth, front to back and front again, as you press the bubbles about the palate. Norman cider, which is often not as dry as Breton cider and usually has less than 4 percent alcohol, bites the throat softly. Brighter and clearer and a point or two stronger, Breton cider is generally sharper, friskier, though each one varies from farm to farm, according to the particular mix of apples and fermentation style each farmer uses. I began to find my cider mouth after the third swallow, a progression I've noticed wherever I've served ciders to the uninitiated. A few take to it on the first taste. Most, however, expecting a variation on wine or beer, express polite confusion or disappointment. But I always make sure to pour a second glass. By the third or fourth swallow, if the cider has any merit at all, they set aside their initial expectations and begin to explore the taste, feel, and aroma more freely. It was just that experience that captured me a quarter century ago in Paris as we easily finished off the entire bottle (at a price of six francs, or a little over a dollar, as I remember) and made our way to the movies, warm spirited but sober.

Two decades passed before I sat down with Étienne Dupont to taste the ciders and aged Calvados that had made him even more famous than his ancestors. The cider he offered me was the richest, most sophisticated I have tasted before or since. When I told him I wanted to learn to make such a cider on our own farm in Kentucky, he offered a paternalistic smile instead of the usual Gallic shrug. "You should try to make what you can make, but it will not be as we make it here. Even if you can grow the same apples and follow the same technique. Because what we make here is on this soil, with this sun and this rain. That you cannot have, and that is why I think we can make the very best *cidre* that can be made."

Pausing only momentarily, he continued, "Whether anyone can be making it in twenty years, thirty years, well, that is another question."

Of course, I took the bait.

"Pernod-Ricard," he answered.

Pernod-Ricard is to the French liquor industry what Seagram's and Heublein are to beverages in North America. Visitors remember Pernod as a logo on sunny café umbrellas or as the licorice-tasting aperitif that turns cloudy with the addition of water. Aside from making aperitifs and liqueurs,

Pernod-Ricard operates a trans-European beverage distribution that includes Coca-Cola and Orangina. Pernod-Ricard has in recent years taken an interest in resuscitating the cider industry. By 1995 Pernod-Ricard or its subsidiaries produced more than 60 percent of the cider made in France, described generally as *cidre industriel.*

The Pernod-Ricard companies have spent millions promoting cider as the lighter, low-alcohol, healthy (containing vitamins C and B) alternative to wine or beer that is good even for youngsters. It has been the single largest sponsor of new research into production technology, and it has paid farmers to grow new, modern high-density orchards to increase France's annual yield of cider apples. Why, given all that, I asked Dupont, did M. Pernod's *cidre industriel* raise such a fuss?

"If you don't mind cider made from Polish apple concentrate . . . ," he answered. "This is not what we do here."

The difference between what Pernod-Ricard does and what the traditional artisanal cider makers of Normandy's Pays d'Auge do addresses the critical dilemma facing the whole of European agriculture as the European Community (EC) struggles to integrate the Continental economy while protecting the peculiar regional character of its farm products. A parallel contest is all but settled in Britain, even as the concentration and industrialization of American agriculture is already a fait accompli. Stated most plainly, the world is awash in food. The French produce more grain, more wine, and more apples than they have any possibility of consuming. Eastern Europeans, particularly the Poles, whose land and labor are cheap, are quickly turning their fields to factory agriculture. The General Agreement on Tariffs and Trade, as well as internal EC agreements, has been designed to remove as many protective barriers and subsidies as possible for farm commodities. As a result, farm survival requires farm efficiency, which means bigger plots, more mechanization, more produce, and lower costs per unit. Agricultural economists at American universities, for example, advised apple growers in 1970 that they could not survive on less than 50 acres of orchard. By 1980, the number of acres had risen to 100, and by 1990, orchards smaller than 150 acres were not considered viable. European countries, and in particular France, have retained farm-support programs to keep small growers on the land. Protective farm policies continue to win general support because the French regard gastronomy, and regional specialties like cheeses, goose pâté, and certain fowl, as the soul of French culture. The price of bread, a triggering factor in the outbreak of the French Revolution, has long been nationally regulated. A law passed by the National Assembly even went so far as to limit the number of supermarkets that could be opened in towns and villages as a way of preserving the bread, butcher, and vegetable shops on which village life depended, but the law was repealed after wholesale corruption rendered it unenforceable. Nonetheless, most French city dwellers, who account for 75 percent of the population, now buy their staples and much of their fresh produce in supermarkets. And right next to the soft drinks and the beer, they will find one-liter plastic bottles filled with cider. It is not a cider Étienne Dupont would care to drink, but a single suburban store in the Paris suburbs probably handles more *cidre industriel* in a week than Dupont sells in three months.

"But is that really cider?" Dupont asks. "You know, there was a definition of French cider. Apples and pears. Fresh cider apples [as distinguished from sweet dessert apples] and some technical points of production.

"In 1987 Pernod-Ricard decided that since France was in the Common Market the French definition of cider had to change to permit the use of [apple] concentrate in order to compete with the English in the EEC [European Economic Community]. To permit the use of table apples from anywhere and the use of concentrate.

"We said, '*Non!*' "

Dupont and his comrades went further, however. They created the Syndicat du Pays d'Auge and began the process of petitioning the Ministry of Agriculture for a precise regulation of what constitutes *cidre du Pays d'Auge*, or as wine makers would call it, an *appellation contrôlée du Pays d'Auge*. France, whose legacy of central, regulatory administration dates to the inauguration of Charlemagne in the year 800, does not grant such petitions easily. Wine makers have never regarded the *cidrières* as *compagnons*, and it was for and by the wine makers that the controlled labeling system was established.

"The world of wine is different from the world of farmers," Dupont explained. The cider men, however, were so incidental to French agriculture that they were not members of the farmers syndicate. "So," he went on, "we created our own syndicate for obtaining the *appellation contrôlée du Pays d'Auge*. Fifty farmers."

Initially they sought only to change French law to block the use of foreign or domestic table apples in any French cider, but understandably the Pernod-Ricard interests fought them off on that point. Petitioning for *contrôlée* status was a far more ambitious strategy, but Pernod-Ricard at least had less direct influence in the wine industry. Their first hurdle was the basic law governing the Institut National d'Appellation d'Origine, or INAO. INAO had jurisdiction over wine and spirits—Calvados has been regulated since 1942—but not over apples. In 1988, the law was amended, and INAO was given money to control cider labeling. That had gone easily. But then the process ground down. The syndicate wanted very precise conditions established that would distinguish the traditional, artisanal ciders made in the Pays d'Auge with Pays d'Auge apples from all others—whether they came from Pernod-Ricard's vats of industrial concentrate or from the artisanal ciders of Brittany.

"We worked, and we waited. Yes, soon," they told us. "Yes. Tomorrow. We want everyone to agree. Well, finally the chairman of the INAO understood there was no way to go against the pressure of Pernod-Ricard."

At this point in his story, Dupont said simply that the Presidential election of 1995 changed the terms. Until then the government of François Mitterrand had been unwilling to move. After the election of Jacques Chirac, the Agriculture Ministry decided to support the partisans of Pays d'Auge (the word *auge* derives from the trough made of a hollowed-out log in which the apples were crushed in medieval times). Oddly, it was Chirac who had long had strong relations with Paul Ricard, the octogenarian patriarch of Pernod-Ricard. What brought the change? On that point, Dupont, who is far from unsophisticated in French politics, had nothing to say. The Pernod-Ricard organization found no reason to talk about the question.

Nonetheless, by February 1996 the farmers had won their battle and the right to use the *appellation contrôlée* signifier on their labels. And what did they gain? The Pernod-Ricard supermarket ciders sold at about seven francs a bottle. Étienne Dupont's sold for between twenty-two and thirty francs a bottle, though his wholesale price netted him less than ten francs per bottle. The cost of growing his apples—more than forty varieties for cider and Calvados—cost him, he said, one franc fifty per bottle. Add labor, equipment, and supplies, and his production costs alone would very nearly equal Pernod-Ricard's retail price. Winning the prestige that came with controlled status was, he believed, the only way he and the other quality cider makers could survive.

Dupont, it seemed, had fought the good fight and won. Tradition, quality, standards, and respect for the land and for the art of agriculture had prevailed against the onslaught of yet another commercial behemoth. And yet I could not quite forget the remark that Joseph Heitz, one of the finest Sonoma County, California, wine makers, had made to me years earlier when I was finishing a report on how sales had flattened out for another mediocre giant, the Gallo brothers wine company.

"You must take some satisfaction in Gallo's troubles," I had said.

A tall hickory stick of a man, he scanned me carefully and in a slight, residual German accent answered, "Listen, sonny, the Gallos made it possible for wine makers like me to do business. The Gallos are the ones who introduced wine to middle-class Americans. If some of them move on to smaller producers like me, it's because they first learned about wine from people like the Gallo brothers. I wouldn't be here without them." America in the 1950s and 1960s and France in the 1990s are, of course, two different places, but there are parallels between the two. Wine was seen by American men as a beverage for either snobs or women. One of the early advertising campaigns for Petri wine featured a brawny woodsman drinking it from a leather flask to encourage real men to feel safe sipping wine in public ("A man named Petri, he's the one, who drinks a good man's wine"). The French don't suffer much gender confusion around their drinking habits, but cider has suffered serious image problems. As much as they may cherish Monet's haystacks and van Gogh's spring orchards, few French people care to be identified with the stooped peasants who tended either. Cider is fine for a weekend in Normandy or lunch at a *crêperie*, but . . . In the quarter century I have been staying in the household of my closest friends in Paris, they have never once brought a bottle of cider to the table.

The numbers capture the cider makers' problem. French consumers drink less than one-half gallon of cider per person per year, as against sixteen gallons of wine. In 1990 cider claimed only 1.2 percent of the $1 billion beverage trade, and while cider sales had begun to rise, they were not keeping pace with general growth in beverage consumption. The one encouraging indicator was exports. French cider sales abroad rose 40 percent in 1991 over 1990. Overwhelmingly, those sales were of *cidre industriel*, the provenance of the Pernod-Ricard enterprises, and the leading importers were Germany (at more than 80 percent), followed by Belgium-Luxembourg, the United Kingdom, Japan, and Switzerland. Combined, all those exports ran about twelve million liters per year in the 1990s—not even a thimbleful in the world alcohol trade. But the exports do indicate a strategy and a target, and it is not the rustic peasant. The new cider drinker is urban, between thirty and fifty, often female, concerned about health, and eager to find a lighter, lower-calorie alternative to wines and liqueurs. Like Guinness (whose slogan announced "Guinness Is Good for You"), the French national cider makers association promotes cider's healthy assets, its vitamins, plus calcium and potassium, all retained through fermentation, and its apparent ability to reduce cholesterol.

At the same time, the association has launched new campaigns to insert cider into modern French cuisine, encouraging food writers to promote the several varieties of French ciders as marinades for meats and accompaniments for cheeses, game, and fish. That attempt to reposition cider from the peasant's bowl to the gourmand's table, of course, benefits the *appellation contrôlée* producers like Étienne Dupont as much as, if not more than, the industrial producers.

"I say that cider should be a vital part of the *menu gastronomique*," Dupont told me, in a near echo of the national campaign. "It is very good to begin the meal, very good with many desserts," and then, more aggressively, he added, "Have you ever tried to have a glass of wine with chèvre, Camembert, Pont l'Evêque?"

He paused, raised his eyebrows subtly, and added a slight shrug. "It is not very good." Then another pause.

"There is a saying, you know. 'If you use cream, you take cider.' I did not know it, but when I read it, I tried it, and it is very intelligent."

Campaigns to reposition cider as the lighter, healthier, more refreshing alcohol with a rustic past face yet another barrier. Even though the flavor and aroma of cider are much less complex than those of a fine wine, cider makers face much more daunting technical problems than wine makers do. Their biggest problem is sugar—or lack of it. Grapes contain roughly twice the amount of fructose that apples do, which means they generally will ferment to twice as much alcohol. Fermentation of wine or cider depends upon the presence of both yeasts and bacteria, but only certain yeasts and bacteria. The higher alcohol levels in fermenting grape juice provide a certain degree of protection against unwanted bacterial and yeast infections that can spoil the wine and create an "off" taste and inconsistency. Because cider seldom exceeds 6 percent alcohol, and spends most of its fermentation time at 2 or 3 percent alcohol, the chance for spoilage is much greater. Even more complex, however, is the question of yeast.

Yeasts, combined with that untranslatable word *terroir* (a combination of geology, soil, sun, rain, and temperature), are at the heart of French gastronomy. The yeasts that make fine ciders are, broadly speaking, the same yeasts that make fine cheeses. Unlike the single yeasts isolated and inoculated into the grape must, cheese and ciders prefer a promiscuous assault of multiple yeasts. There are at least four distinct yeasts, each with scores of strains or variations, and five separate bacteria required to make a fine cider. Or, as Jean-François Drilleau expressed it: "Cider is not the result of fermenting sterile apple juice inoculated by a single yeast strain."

Drilleau, I had been told, runs the only laboratory in the world fully dedicated to understanding the microbiology of fermented cider. A parallel lab in England died for lack of funding during the Thatcher years. Even at Drilleau's lab, outside Rennes, at the eastern edge of Brittany, there are only four scientists and technicians.

Drilleau, fiftyish and graying, was as modest as Étienne Dupont was grand. He met me at the drably efficient train station in Rennes, a city rebuilt by witless postwar architects who managed to eliminate whatever urban charm the Germans hadn't already destroyed. We had hardly reached the edge of the city when Drilleau framed his dilemma: "In Europe we have two kinds of cider making—the way they do it in England, which is only one-third apple juice concentrate mixed with corn syrup, citric acid, and coloring plus carbon dioxide gas, and the way we do it in France where we try to make the cider through more traditional ways. But I do not know if we use the right way." He shifted the gear on his small, government-issue Simca and looked over to me. "We do not make so much money."

The cider scientist's remark was as political as it was technical, for Drilleau and his team operate within the vortex of the traditionalists and the industrialists. Traditional cider making relies not only on specific types of cider apples, most of which are small and astringent, a result of the tannin that gives body to both wine and cider, but also on wild yeasts and bacteria for fermentation. Those microflora are ever present on the ground, in the air, on the skins of the apples, and in the grinding and pressing equipment that turns the apples into juice. The complement of each single yeast or bacterium varies from farm to farm and, to some degree, from year to year, as heat and rainfall vary. Thus, like wine, ciders change from one season, or vintage, to the next. Raising further complications, the fizz in French cider comes from a final, natural fermentation inside the bottle that gradually uses up residual sugars and raises the percentage of alcohol, making the cider much drier after a year than it is at bottling time. Like the lambic beers from Belgium, the cider is "alive" and as a result somewhat volatile. The surest sign of such traditional ciders is the thin layer of yeast sediment at the bottom of the bottle, which will turn the bottle cloudy if you shake it (a practice the Asturians in Spain always follow and one that shocks and offends the French).

Reliance on wild yeasts does two things. It leaves a distinctive taste print on each producer's cider. It also throws consistency to the wind. The fact that alcohol levels vary over time creates other headaches. Wine makers solve the first problem by adding sodium or potassium sulfites that suppress the wild yeasts; then they add cultured, commercial yeasts, whose effects are consistent year after year. Champagne makers ferment their wine to a desired dryness, then filter out the original yeasts and add a *dosage* of sugar or fructose and champagne yeasts to build bubbles inside the bottle. Up to that point their technique is not so different from that of the cider makers, but then they undertake a process called *dégorgement*, in which they turn and tilt the neck of the bottle downward so that the new sediment collects at the mouth, where it is frozen. The bottle is re-opened, an icy plug of sediment is removed, and the cold champagne is quickly topped off and re-corked to hold the carbon dioxide inside. A few cider makers have tried this so-called *méthode champagnoise* to clarify and stabilize their ciders, but the cost of all the turning and chilling required for *dégorgement* prices the cider beyond customer tolerance. Instead, traditional cider makers argue that their drink's "aliveness" and its variations are marks of its artisanal heritage. That was fine, so long as the cider customer was a farmer who lived no farther than a few kilometers away and each producer bottled only a few hundred to a few thousand cases. The new urban consumers of Germany, Japan, or even Paris, who have been targeted as cider's salvation, may not, however, be so forgiving. Still more problematical: should a container of "live" cider cases heat up much beyond seventy degrees, the yeasts within might easily reawaken and send the bottles popping all over the highway or the storage warehouse.

Jean-François Drilleau's work, half of which is funded by the government and half by private contracts (most of them from Pernod-Ricard), is to solve these problems, to discover how to make an artisanal style of cider that meets the standards of consistency required by the European market and the global market. "When I came here almost twenty years ago, there was a gap between the old and the new cider makers. Most of the new ones, we had to teach them how to make good cider. But as we don't know, we can only tell them when they have a bad cider. We don't know, scientifically, how you make a *good* one." A key example is a process the English call "keeving" and the French call *défécation*. After the apples were pressed, a small amount of salt and wood ashes were added to the juice. If it was cold enough (no more than fifty degrees), and if the farmer hadn't allowed too much manure into the orchard to raise the nitrogen level in the apples too high, the apple must would become much clearer. In about a week, starch and other sediment would sink to the bottom of the vat; pectins would rise to the top and form a thick, gelatinous cover called a *chapeau brun*, or brown cap.

No one knew how *défécation* had arisen, and initially no one even knew exactly why. "We discovered," Drilleau explained, "that it was used to take off the phenyl content of the juice so you can have a slower fermentation in a clear medium instead of a fast, turbid cloudy process. The wood ashes contained a lot of calcium that helped form the *chapeau brun* and reduce the biomass of the juice. Now we tell the cider makers to add specific amounts of calcium chloride and pectinase." The *défécation* served two purposes. First, it reduced the nutrients that fed the more aggressive Saccharomyces yeast strains (the primary yeast family used in wine making) and thereby allowed the other wild yeast strains time to grow in the clarified must. Second, by slowing down the fermentation, more of the vital sugars were preserved, which in turn allowed flavor-controlling esters to develop during the succeeding four to six months of maturation.

Some industrial cider plants still rely on *défécation*. Others run the juice through a centrifuge to clarify it, a process the traditionals believe ruins both flavor and aroma. In itself, however, *défécation* was essential for a beverage made with a fragile mix of wild yeasts that would yield the kind of complexity that defined true cider.

Drilleau walked over to a cabinet in his lab and pulled out a large spiral notebook containing several acetate slides. On the first a green line formed a nine-sided, wobbly circle, like a corral built by drunken cowboys. At the 12 o'clock position was the word "color"; at 2 o'clock, "aroma"; at 3, "green apple" for unripe; at 5, "fruitiness"; at 6, "floral"; at 7, "acidity"; at 9, "mold"; at 11, "*fermier*," or "barnyard." The further each characteristic was from the center, the more of that quality showed up in the cider. "We try to reproduce a traditional cider taste by testing and using new yeasts that we can now grow. Each one will have a different figure on the graph to show its qualities. It was twenty years ago that we started to identify the different strains of yeast. So if we use another one, we get a different-shaped figure. If you want a cider that is more floral—a more flowery bouquet—you can play with another yeast."

He pulled out another acetate figure drawn in red.

"This one, you see, gives high aroma, but it has high *fermier*, which we might say is a mix of straw and manure—not so good—and low fruit." Another sheet showed a blend of yeasts. Not only the yeast strains affect flavor. The time at which they are added to the cider matters. Some may survive for only a few hours midway through fermentation but still affect the color and texture of the cider. Other yeasts battle one another, slowing or speeding the fermentation, depending on how far the sugar-alcohol conversion has proceeded.

When I asked to make copies of these yeast graphs, Drilleau shrugged, explaining that the exact details were privileged property, paid for by one of the research station's large private contractors. So far, after twenty years of primary research, no clients, no cider makers have yet been able to develop and use a complex mixed-yeast package that will produce an acceptable traditional-tasting cider. Many avoid the problem of volatile yeasts working inside the bottle by filtering all the yeast out and pumping carbon dioxide in to create effervescence. After tasting several unmarked ciders from the lab cooler, I had to acknowledge that the carbon-dioxide-infused one was better than others I had tasted with naturally created bubbles. Yet the main problem producers face persists: how to create a genuine

French cider character but maintain the consistency that distant markets require.

The patrimony of French cider is as old as France. Charlemagne, the first Carolingian ruler of the Franks, directed peasants to plant cider orchards in the year 800, halfway through his reign. Ciders seem to have been drunk by the Romans, though they may or may not have been drawn exclusively from apples. Modern cider apples (meaning those of the last thousand years) likely arrived in Normandy from trading missions with the Asturians and Basques in northern Spain, both of whom claim their ciders predate the Roman conquest. Farmers and farm villagers treated cider, which was fermented in barrels and tapped by draft, like water, pouring it freely for small children and elderly grandparents, supplying three bottles a day, morning, noon, and evening, to peasant laborers in the field. While the Celts probably fermented wild forest crab apples in Britain even before the arrival of the Romans, it was the Normans who seriously introduced cider making to England after the conquest of 1066.

Three centuries later, cider was fast emerging as the preferred beverage of the English (thanks in no small part to the blockage of wine from France during the Hundred Years War). In France only the Normans and the Bretons had succeeded in replacing wine with cider—largely because their wine was barely potable. But among the English, apples ruled supreme. The soft breezes of the Gulf Stream warmed all the West Country—Somerset, Cornwall, Devon, Wales, and Hereford

—making apples as easy to grow as grass. Sweet dessert apples were planted to the east in Kent, but it was in the west counties that the famous bittersharps and bittersweets soon covered thousands of rolling acres, bearing fanciful names like Brown Snout, Foxwhelp, Slack-Me-Girdle, and Tom Putt. Easier to grow than sweet apples because appearance was of no matter, cider apples tended themselves and became an integral element of every farm.

Broad historical forces favored cider as well. By the seventeenth century the enclosure of land had reached feverish proportions. More and more peasants were either driven off the land and into industrial mills or forced to work for the great estates as chattel serfs. The estate owners themselves found that they were being drawn deeper into the world's first industrial economy, that the cycles of supply and demand in the cities had direct consequences on the land. At first, industrialization and urbanization favored cider over other drinks. Perennial wood shortages led local authorities to urge planting of apple and pear trees, which could eventually be recycled for fuel or as knee joints in shipbuilding. More important, brewing grains into beer and ale required great quantities of wood, much of which had to be imported, to cook the malt. Apples fermented on their own, and thus, in Britain's constant tension with the Continent, cider became the patriotic beverage of self-sufficiency. Then, in the 1660s, after the Restoration, agriculture fell into a deep depression. Neither grain nor fresh fruit found a strong market. Farmers turned still more of their land to cheap cider trees that needed no tending and could be easily integrated with the expanding herds of sheep that grazed beneath them. Wool, which after all had fueled the enclosure movement and set off the Industrial Revolution, was constantly in demand in the cities, and the pomace left after the autumn apple pressing supplied free feed to sheep, cattle, and horses.

Cider played another role among the gentry and nobility who inhabited the great estate houses. They began to think of their *cyders* as the French dukes and counts thought of their wines. Fine ciders became emblems of fine living—in an English mode. Propelled by the fascination with rationalism and natural science, a new *scientific* horticulture spread among the estates, especially in Herefordshire. Its great proponent John Worlidge maintained that just as fine wines could be produced only from particular grapes fine ciders required the study, identification, and propagation of only the finest cider apples. His *Vinetum Britannicum, or, A Treatise of Cider etc.* (1676), in which he catalogued all the known cider apples and evaluated each for its strengths and weaknesses, became the first horticultural bible. Two centuries would pass before chemists learned how to make precise measurements of the amount of potential alcohol each apple contained based on its sugar content. Worlidge and other cider specialists set about fermenting apple varieties individually to evaluate strength and savor by taste, just as Continental wine makers had done for centuries. Special panels of judges conducted tastings in which they pitted cider labels against one another and even against fine French wines. Members of the Royal Society debated agricultural procedures and fermentation techniques. Their object was unabashed: to displace grape-based wines with the ciders that they believed would eventually prove superior to even the finest Burgundies. Two of the most favored Hereford varieties, Redstreak and Foxwhelp, possess such high levels of sugars that in a good year they can ferment to 11 percent alcohol, about the same as a light Chablis or a Moselle. Golden Harvey, or Brandy Apple, could reach 12 percent.

These last years of the seventeenth century were cider's flower. England controlled the seas uncontested. Her triumphs in science, architecture, shipping, and commerce led her ruling families to believe that they now defined the very standards of civilization. The new specialty ciders offered them proof that science married to tradition could lead to unsurpassed levels of refinement. Development of new, sturdy glass bottles, which even the French lacked, enabled cider masters to capture the final fermentation inside the bottle, yielding a long-lasting fizzy apple champagne that could be shipped and poured at the most stylish parties in London and Bristol, where elegant gentlemen confirmed cider's superiority. Worlidge, triumphant, wrote, "I hope every English man; or Native of this Isle on his return hither will conclude with me, that our British Fruits yield us the best beverages, and of these Fruits the Apple the best, which is here called Cider." John Beale, a friend of Worlidge's and a prominent Hereford parson, declared: "Few cottages, Yea, very few of our wealthiest yeomen, do taste any other drink in the family, except at some special festivals, and that for variety, rather than for choice. And when the King (of blessed memory) came to Hereford, . . . both King and Nobility and Gentry did prefer Cider before the best Wines those parts afforded." Doctors prescribed cider as a curative for gout, rheumatism, and urinary infections, and surgeons used it to cleanse wounds. Captain James Cook even took cider with him to prevent scurvy among his crew on his 1776 voyage to the South Seas.

At the end of the seventeenth century England was awash in cider. Yet by the end of the next century, farmers were hitching oxen to their apple trees and ripping them out, and rich Burgundies were once again filling the goblets of the finest dining rooms. Commerce, more than anything, had burst the cider bubble. As busy as Worlidge and his horticultural colleagues had been blending fine bottled ciders, the commercial vendors had been busier selling weak common ciders to the poor in the public houses. As volumes rose, prices rode the predictable roller coaster of boom and bust, and quality suffered. Simple farm families had long diluted their ciders with water or wet down the residual pomace for a second, weaker pressing. But the brokers soon began to make their own dilute blends. Production records dug up by English cider historian R. K. French suggest that farmers were pressing about twice as much cider per acre of apples at the end of the eighteenth century as at the beginning, which could only have been accomplished by watering it down. Water, alas, was not all they added. Other fruit, sound or rotten, other juices, sugars, even sheep's heads and entrails—anything that would ferment— found their way into the huge wooden vats. All of it, eventually falling under the term "scrumpy," was blended into the rough, fiery apple brews favored by sailors along the docks of Bristol well into the twentieth century.

Then came the gin wars. The Dutch, who had become the consummate commercial shippers and traders of Europe, were looking for new, cheap products to complement wines. Gin, their own distillate of grains and juniper berries, was compact, easy to transport and cheap enough for the commoners. As gin made its way up the Thames, the native brewers fought back, drawing moralists and health advocates to their cause. Gin was called a "mother's ruin," while beer and ale were healthy native alternatives. By the time the battle was done, the three great brewing combines, Bass, Guinness, and Whitbread, had firmly implanted themselves in the English drinking diet, steadily organizing the licensing system that even today controls most of British pub life. Cider had failed to compete. Worse still, the prime minister, the Scottish Lord Bute, had imposed a separate excise tax on cider equal to the tax on wine. The tax didn't last long, but combined with the new appetite for hops, and recurring scandals over the degraded quality of popular cider, the fermented apple began its long slide downward as nothing more than a poor man's quaff. "That produced in the London Taverns . . . bears so little affinity to prime cider," a traveler wrote in the early 1800s, ". . . that, when tasted by a person conversant in these, it appears no other thing than a whirligig composition of such vapouring elements as he can never hope to reconcile or compose."

Today, on the edge of Hereford, in the West Midlands, there stand a dozen 150-foot-tall steel tanks that hold 1.6 million gallons of cider each, surrounded by a virtual village of smaller vats designed to hold, altogether, 15 million gallons of cider. Each tank has its own name: Pluto, Mars, Venus, Asia, Australia, Wren, Robin, and so on. This is H. P. Bulmer Ltd., the largest, most prosperous cider company in the world. The first of those 1.6-million-gallon tanks was put up in 1975. It was named Strongbow, the brand name for the product that the

Bulmer family believes will drive a worldwide restoration of cider and create the first global cider market. George Thomas, a bushy-browed cider maven who speaks for the Bulmer company, laid out the cider strategy: "Bulmers has sold cider in dozens of countries around the world but never with a great strategy or a grand plan to create a 'global market' like our friends Coca-Cola did all those years ago. You have Coca-Cola wherever you go; whether it is Moscow or some little, abandoned island, you will find a Coca-Cola machine, and it will be the same Coca-Cola everywhere, and the same image, and the same branding, and so there is a market and a grand plan.

"We are now doing that for Strongbow. Strongbow is now on test, on draft in America, in several European countries, in Australia. We have a long-term plan that will eventually create a new world cider market, led of course by Strongbow."

Strongbow is only one of seventeen cider labels Bulmers makes. Until Strongbow came along in the 1960s, Woodpecker, a sweeter cider at 3.5 percent alcohol (Strongbow is 4.5), claimed the strongest market, and in America it is still the leader. Others include Bulmers Original at 6 percent; Strongbow Ice, smoother and lighter but slightly stronger at 6.5 percent; White Lightning, filtered clear "white" and much stronger at 8.4 percent; and Old Hazy, a cloudy, naturally effervescent cider, traditionally fermented, at 5.5 percent. The company's fortune, however, has all been bet on Strongbow, which, intriguingly, nowhere includes the word "cider" on its label. Instead, it bears a fierce, bronze-helmeted medieval bowman, the string of his bow drawn tight.

Like the French industry, English cider was failing fast through the 1980s, down to seventy million gallons. Small craft cider makers were closing in Somerset and the southwest, and Bulmers itself was losing sales. By 1988 the family and the corporation, which had gone public in 1970, realized even they were not strong enough to survive the steadily falling market. The family, who still owned more than half the stock and held the top jobs in the company, took a gamble. They launched a completely new marketing strategy, more than doubling spending from five million pounds to eleven million.

"That immediately resulted in a five-million-pound drop on our profits," Thomas said. "Our shares collapsed overnight because the City, the stock market, are short-term thinkers. Now the Bulmer family had committed themselves to that bold strategy that had been presented to them by the marketing team. The family said yes . . . If it hadn't have worked . . . well, what the heck, we were going downhill anyway.

"But it did work, and today the family have been repaid handsomely and so has the City, because our shares are now blockbusters. Today the Bulmer family are among the richest in England from their success as cider makers. The company has a turnover of almost 260 million pounds a year, profits of 27 million."

By 1996, seven years after what the Bulmer people refer to as "the regeneration," British cider sales had risen to 120 million gallons, more than half of them Bulmers' beverages. Moreover, they boast, much of that growth came during the worst recession since the Great Depression, and they believe the national market will top 150 million gallons by the millennium.

Again, and as in France, traditional cider partisans ask, "But is it cider?" Some, at least, respond as old Joseph Heitz did when he explained to me the importance of Gallo jug wine.

Westons is an old, regional cider maker just down the road from Bulmers. Weston cider is "cask fermented," meaning that the last fermentation takes place in a pressurized keg and is pumped directly to bar taps. A pale amber, very lightly bubbly, and with a slightly musky back taste, it is never refrigerated. "The managing director there used to be a friend of mine," George Thomas went on, "and he used to ring me up every New Year and say, 'How much are Bulmers spending on advertising this year, George?' And I'd say, 'Twenty-five million.' 'Oh, jolly good, that means I don't have to spend a penny,' he'd answer. Because the success that we were generating was percolating down to all the others and so people were becoming more and more interested in cider, and as you become more interested in cider, you are looking for different styles of cider, the kind of styles of cider that Julian Temperley makes down in Somerset, for example."

Crusty, moody, contrarian, cantankerous, Julian Temperley is the brilliant bad boy of English cider, a man constitutionally incapable of offering a kind word about the H. P. Bulmer company or any of his other competitors. His orchards carpet the low, flat fields known as the Somerset Levels, down west, beyond Arthurian Glastonbury, where the ancient roads have sunk into winding green ditches, barely wider than a wagon, hedge tunnels in the night, marked by signs pointing to places called Othery and Mallet, Muchelney and Midelney, Yeovil and Eel Smoker, Crewkerne, Combwich, and Viking Pill.

To find Julian Temperley's place, everyone told us, look for the solitary hill with a lone oak on the top, Burrow Hill. His place was just across the road, a mile or so from Kingsbury Episcopi. Somerset folk give directions the way country people everywhere do, not by names or numbers of roads, but by visual landmarks—a stone barn, a crooked oak, a set of thatched cottages. I passed a newly thatched house, but it was too grand, and the BMW out front too expensive, to be a farmer's cottage. The lone tree stood high to my left beyond the highway hedge. A small wooden sign on the right read PASS VAIL FARM. I turned the car into a short drive between an old, oak-timbered barn and a genuine country house.

Leaping sheepdogs barked and threw their paws against my driver's window, eager only to have their ears rubbed. Straight ahead stretched an orchard full of old-fashioned, double-grafted apple trees, limbs branching out four feet above the ground, high enough for sheep and hissing geese to wander about. To the left, spread out on the cement, lay great piles of apples, red, gold, and green, six, eight, ten feet high. Tall, yellowing fiberglass fermentation tanks stood back behind the apples and the barn that sheltered a cider press in its inner shadows. A well-groomed black Vietnamese sow sauntered low to the ground policing the premises for sweet favors. Temperley was nowhere about.

An hour later he appeared with his wife and daughter, on a buckboard buggy pulled by a clumpy white draft horse. It was Sunday, and they'd been out for an afternoon ride. Temperley makes a tall, lumbering figure, modestly paunched and deliberate of stride. He was pleased to talk and offer me coffee and tastes of his Burrow Hill Cider, drawn fresh from the barrel. His was the first traditional West Country cider I had ever sipped. Absolutely flat, always taken at room temperature, it smells and tastes of an odd, greenish musk behind the fruit. Almost all of it is sold right there in the barn to local people whose ancestors have been drinking it that way for several centuries. He pumps it out of great oak vats, each with its own name rather than a number—Big Fat Vat, Huge Vat, Newton Abbot Vat—into smaller barrels and then into bottles, jugs, or two-pound plastic casks.

At first I thought it might make a fine vinegar. But I bought several bottles anyway, and over the following weeks, the cider and I developed a respect for each other. Temperley's brandy, however, was the thing earning him headlines in the London papers. "Tipple that's the toast of Somerset," *The Independent* said, followed by, "Kathryn McWhirter [the author] finds a cider brandy from the West Country to be the equal, perhaps the superior, of its finest Calvados cousins." Reports like these had brought minor fame to the onetime student agitator turned cider master and distiller. A fancy wine and brandy purveyor with a shop just beside Windsor Castle showed up the same afternoon to cart off several cases, but only after tasting a few samples.

Calvados is famed for warming up shivering souls in the rawest, wettest weather. Usually it has been aged in oak for half a dozen years and best of all a dozen and a half. Étienne Dupont, better known for his "Calva" than for his cider, sells some that has been in the barrel for more than forty years. Temperley's Somerset Cider Brandy was only two years old that fall of 1992, and it swallowed like a fire in the throat. Four years later, on a drippy December Sunday, I stood with friends in the same chilly barn and tasted it again. Now it was richer, fuller flavored though still bright with the aroma of apples, and, as one of my fellow travelers said, it poured all the color of autumn into the glass.

Somerset Cider Brandy is the only commercial apple brandy distilled in England and has survived only because Temperley successfully fought off a Common Market challenge from the Spanish, who argued that brandy could not be an English product. Calvados could only be made in Calvados, cognac in Cognac, and brandies in either France or Spain, they claimed. Temperley and his partners, however, had already invested one million pounds in Somerset Cider Brandy, and they were hardly ready to fall into their stills. They might have expected national solidarity to draw the Scotch whiskey industry to their side, but the whiskey association's protective instincts won out, and it joined the Spanish effort to close Somerset down.

"We had problems with the Spanish and the Scotch Whiskey Association," he told me, his sentences trailing off into a characteristic, clipped mumble. "Spanish bureaucrat, really, trying to . . . reading the regulations wrong." Then a snort. "And the Scotch whiskey people asked our government to support the Spanish."

Suspicion of creeping mediocrity brought on by the accountancies of the growing global market lurked just beneath everything Julian Temperley had to say. He was rather fond of the Scots themselves, the individual distillers and coopers who taught him about barrel maintenance and aging techniques. "I don't think it was actually the Scots, but all the Scotch whiskey industry is owned by two, three, four companies, multinational companies. They look to protecting themselves against anything that they see as a threat. They were protective of the fact that in places like Fortnum and Mason, in the quality-spirits section, there is now cider brandy alongside of the malt whiskey, and they look at every sale of cider brandy as a sale off malt whiskey.

"In some ways it is an enormous accolade to get into their bad books, but it is depressing in a way. The Whiskey Association is actually run by Englishmen, and they are bureaucrats doing a job. But they are wrong." A droll smile spread across his face and gave a chuckle to his broad, funky farmer's frame. "And they lost, which is even more pleasing."

He pulled off an exceptionally handsome bottle of Somerset Cider Brandy. "Our aim is to produce a quality spirit in the West Country that will sit on the shelves of Fortnum and Mason, and this bottle is exclusive to Fortnum and Mason in London." He handed me the bottle, priced at about eighty dollars.

"With an apple in the bottle, as you see, and if we've caused the Scots any grief, then we must be doing something right."

As for cider, the sort of farmhouse beverage that has sustained the farmers of Hereford and Devon and Somerset for centuries and which until recently was their principal income, Temperley was pessimistic. Pessimistic about everything to do with it. About the new dwarfing (or as the English say "bush") trees being planted, which he swore couldn't produce a good cider apple. About the eagerness to spray the orchards with pesticides. About the use of rotten apples and concentrate and commercial yeasts and carbon dioxide bubbles. "If the ultimate place of a cider is for somebody to get legless drunk, they don't particularly mind whether it happens to be a rotten apple. That is actually where the English cider industry has gone to, over the last ten or fifteen years. Supplying cheap alcohol to hooligans and unemployed and sick and the housewives. So there is no real need there actually . . ." His indictment trailed off, but it kept coming back as a pervasive betrayal of the genuine English character, genuine English heritage, the genuine yeoman, the romantic craftsman, who once, in the murky past, traipsed out into the orchard, banging on the grandest apple tree, feeding wassail to the wren-boy perched high up in the limbs with his offering toast.

"Most of the original cider makers were cider romantics. They believed in cider as something. Now, the difference is that cider is just run by the City accountants. That happens with all businesses. I expect Bill Gates is interested in Microsoft, and is fascinated by computers, but when he goes it might well be taken over by accountants. All the same, it is nice to produce a cider which you can give to fathers-in-law Sunday lunchtime."

Still and all, I asked him, didn't he believe the Bulmer campaigns had helped sell his ciders.

"I think . . ." He stopped. "I don't think it actually bears any relationship to anything." A man dressed in a windbreaker walked into the barn, carrying a tablet, and started to speak.

Temperley clearly disliked the man, a neighbor, who was bearing a survey on a village matter.

"No, I told you not to come here, we are not doing that thing. Will you please leave us alone and stop terrorizing us about . . ."

He grew genuinely angry.

"Will you please leave these premises!"

He picked up a green garden hose, opened the tap, and turned it on the neighbor.

"I'm not putting up with that damn thing."

The man was wet now, starting to leave, but Temperley kept after him with the hose.

"You shouldn't have done that, Julian. You'll be sorry you did that."

"Go! Oh, get out. You are causing so much grief in this village, you buggers. Just go."

And he did. Then Temperley turned to us.

"They just go on and on. We have a village, an old-fashioned village, but there have been a whole lot of people moved in. We have what you call white settlers." He mumbled a bit more about the survey the man was conducting, and he said, "I shouldn't have done that. No, I shouldn't have done that."

One of those villagers, James Crowden, works seasonally at Temperley's. Sometimes he presses the odd little apples into juice. Other times he works with Josephine and Fifi, the two French stills Julian Temperley bought from a farmer in Calvados country in Normandy. (Only apple brandy made there can bear the name "Calvados.") A sheep shearer and a poet as well, Crowden has watched the steady disappearance of the local cider men and their farm orchards.

"Orchards are a great asset and some sheep shearers used to drive into a farm, test the cider, and if it wasn't any good drive on out again, without even looking at the sheep," he wrote in a note to a 1996 collection of his work, *In Time of Flood*. "Fewer and fewer farmers have the time and inclination to make their own cider and regard orchards as an inconvenience rather than as a necessity. White yuppy designer cider however is not good for sheep shearers, though the bottles can be used for drenching sheep in emergencies."

Crowden is a great celebrator of Somerset tradition, of the

stonecutters and eel fishermen, and, not least, of the ornery, contrarian style that has seemed to mark all the independent cider makers. Few ever have betrayed much affection for the government or the Crown, distant powers seen mostly as greedy tax collectors. Under the reign of the demented and much reviled George III, Parliament placed special taxes on cider brandy, showing it to have been an attractive revenue source, but within a hundred years, under Victoria's rule, apple brandy had disappeared, whether from the burden of taxes, change in taste, or the hyper-depopulation of the countryside in the wake of industrialization.

For Crowden as for Julian Temperley cider stands as a sort of spiritual barometer of what is true, honest, and honorable in the English spirit. Crowden wrote evocatively of that spirit in a short poem he called "Burrow Hill Cider Farm."

There in the barn dark	*layer by layer*
Inch by inch	*gallon by gallon*
The steady inquisition	*begins*
The confession extracted	*bit by bit*
As the press pushes home	*life blood of the heathen apple*
Flows ever more freely	*the stout cheese dripping*
Ruddy Brown and golden	*like honey*
A river in flood	*a hive of fruit*
Whole Orchards	*pulped and crushed*
The never ending tide	*trailers*
Ebbing and flowing	*acres deep and rounded*
Mounds of apples	*tipped and spewed forth*
Like long barrows	*the farmyard filled to bursting*
Lagoons of red and yellow . . .	*Brown Snout and Chisel Jersey*

Dabinett and Porter's Perfection   *Stoke Red, Kingston Black*
Bloody Turk, Yarlington Mill   *Lambrook Pippin*
Tremlett's Bitter, Tom Putt   *Somerset Royal*
Their ransomed juice   *pumped to distant vats*
Vast in their yeast brooding   *potent and powerful*
Broad in the beam   *fecund and fattening*
Their froth fermenting bellies   *bound with oak*
Straps of iron, barrels   *giant and gargantuan . . .*
A Norman trick   *this drink from apples*
Distilled and fiery   *hint of orchard on the tongue . . .*

The fecund floods of the Levels course through the history of Somerset memory, waters that drown out the land, fertilize and refreshen it, set it apart from the faded industrial empire that all but forgot its Celtic ancestors. Those brackish waters are not the only floods of Somerset, James Crowden writes. Alongside them in practice and memory run eels, "elvers," "grockles," and "the lifeblood of the heathen apple," all visceral markers of lives built on land and water soon to be washed aside by the tides of the irresistible global market. England, the first and still most industrial of all European nations, has lost its palate for the rough, raw art of artisanal cider, Julian Temperley said as he stood there in the barn beside his intemperate water hose.

"I think . . . ," he answered as I directed him back to the possible beneficial effects of Bulmers and its global Strongbow, "I think that if it is true, then it is very sad . . . that we are just sort of parasites living on somebody else's industry. I don't think it is true, and I hope it is not true. You know they say that old Bertrand Bulmer [the grandfather of the current proprietors] announced that he would never go to drink a Bulmers product."

The visit from the surveyor. The house down the road up for sale to yet another Londoner. The notion that the "hooligan brews" were the very things that were saving the hides of artisans like himself. Julian Temperley's mood grew as dark and melancholy as the wet December clouds. "You take it rationally, and it is an appalling judgment if that is all it is, if that is all we are . . ."

SEVEN

Pomona's Prospect

Mitchell Lynd, who twice in fifteen years appeared on the cover of *American Fruit Grower* as "Orchardist of the Year," had a story to tell me. It was about some Kazakh apple seedlings and cold weather. Cold weather is what brought me to the Lynd orchard. An April freeze had killed more than 80 percent of our apple blossoms in 1997. We needed extra fruit to keep our sales up, and his famous orchard, the largest in Ohio, was dead center of the state, about three and a half hours north of ours. That same spring cold snap had taken half of the Lynd brothers crop, and that was part of his story. As it turned out, the story recalled Ohio's most famous apple man, John Chapman, known to most Americans as Johnny Appleseed, but that came later.

Like every apple grower I've ever met, Mitch wanted to

show me his orchard. It was seven o'clock. Light was falling fast, and supper almost ready, as we headed for his weathered Cadillac to bounce through the orchard, passing silhouetted rows of Galas and Romes and Red Delicious, some odd Asian pear apples, and a scattering of experimental projects. When all we could see was the fruit caught in the beam of his headlights, we gave up until morning, when, he said, he would show me something very special.

Mitch Lynd, I knew already, was a quirky sort of fellow. We'd bumped into each other at a number of fruit meetings. At one of them, the annual gathering of the International Dwarf Fruit Tree Association, generally considered the most technically sophisticated apple meeting of the year, Mitch watched a hotel hallway fill at the end of a panel on sunlight and tree training. "You know, I'll bet you can't find a single conversation here about how to grow apples that taste good."

"Is that how you've grown so successful?" I asked him. "Planting just those apples that taste good?"

"Nope," he answered. "We've got Red and Golden Delicious and Rome Beauties. They pay the bills and they've educated the children. Especially the Romes. They may not have a lot of flavor, but they're totally reliable. We get a crop every year, and we sell every one of them."

On this visit to his home and orchard he wanted to show me his passions. One of them is propagating unusual daylilies. Another is preserving a hundred or so "antique" apple varieties. Several years ago he planted half an acre of Golden Russets. He was sure that this greatly loved eighteenth-century apple could be brought back as a modern consumer favorite. They had grown seven years in the field and were just at the cusp of production when a dozer operator mistakenly tore them out of the ground. Now he had a new passion.

The next morning, after we'd toured through two of his three orchards, we pulled up beside the packing shed. From this building the Lynds had shipped boxes of Romes to Norway, Red Delicious to Saudi Arabia, and Empires to Great Britain. What he wanted to show me, however, was outside the building. Over on the side, protected from the burning southern and western sun, sat 557 plastic nursery pots. Each one contained an apple seedling between ten inches and two feet tall. These infant apple trees, he told me, just might contain the answer to fighting off freezes like the one that stole most of his and our crops the previous April.

The trees were Kazakh seedlings, about half of them from seeds gathered by the USDA's Phil Forsline and his Cornell research team in 1993. That was the team's second visit to Kazakhstan, and it almost proved a disaster. The Kazakhs, always eager for Western contact, had invited Forsline back that year, but they failed to tell him that a spring freeze had killed nearly all the wild forest apples. As he climbed up the mountain ravines, Forsline found almost nothing but barren trees. To make the best of a bad situation, he collected what he could find, supposing that at least these scattered fruit-bearing trees had demonstrated resistance to spring frost.

Forsline brought several buckets of seeds back home and stored them until autumn 1996. He planted some at Geneva and sent two other batches to apple breeders at the University of Minnesota and the University of Arkansas. Apple seeds don't just grow. They must be dried to protect them from rot, but not dried too much. Then they must be chilled (though not too much) before they are able to germinate. Curt Rom, who runs the small fruit-breeding program at Arkansas, set to work quickly, adding the Kazakh seeds to several of his own. But after the chilling was finished, something seemed very wrong. "Almost all of our own crosses, 95 percent, germinated," he told me, "but hardly any of the Kazakh seeds, less than 7 percent, germinated. Of the two thousand seeds Phil sent us, we got thirty-five plants. Very discouraging. So we threw out the rest." Then reports came in from Minnesota: none of those seeds had sprouted either. No one had any explanation, unless the seeds had perhaps been frozen in the cargo hold of the plane coming back from Asia. Or maybe some strange pollinization effect had rendered them sterile, which didn't seem likely.

Then in March there came an E-mail message from Forsline in Geneva: "Hey guys, we have 100 percent emergence!" Only one thing distinguished the Geneva procedure from what the technicians had done in Arkansas and Minnesota. Standard breeding procedure is to chill the seeds for 90 days, after which the seed skins break and send out what is called a "radical," rather like a fetal root, and the seed is set in soil. Phil Forsline had left his Kazakh seeds under refrigeration for 180 days. Rom began searching the literature about seed-chilling periods and germination until he came across an old master's degree thesis from New Jersey titled, "Seed Stratification Correlates to Bloom Date in Apples." In plain language: varieties that bloom late tend to create seeds that germinate late, that require longer chilling time.

"As soon as I read that I called Phil back," Rom said. "All of a sudden, the pieces of the puzzle came together. We threw away the seeds we needed to keep and kept the ones we didn't! Wow! What a find!"

Forsline, fortunately, had germinated about a thousand seedlings, all of which the lab inoculated with fire blight bacteria, scab fungus, and other screening tests. Of those, 207 were selected. The germ-plasm repository at Geneva, however, was not an apple-breeding center, and Forsline had to find a home for his Kazakh progeny. Rom was eager for a second chance. Forsline boxed up the young plants in late June 1997 and shipped them to Fayetteville, Arkansas. No one expected these seedlings to produce good apples, but if they contained genetic material that could be used to create trees that were resistant to cold springs, they would prove immensely valuable. From Missouri and Arkansas all the way to Norfolk, apple growers are always praying they will survive unpredictable late frosts and freezes.

The boxes arrived at the beginning of July—already very late for transplanting seedlings in the South. "The day they arrived, I was working up the ground," Rom told me when I called him at the Arkansas experiment station. "I had the nursery rows all ready. Then the boss blew his gasket. I hadn't put in the appropriate memos to the farm committee. I hadn't followed every procedure. But it was late and we had to get these trees in the ground. I suppose he may have thought it was going to take up twenty acres, but actually they would only have taken around fifty feet in the nursery row." Besides memos, procedures, and review committees, Rom faced far bigger barriers. Apple growing had ceased to be a strong force in Arkansas agriculture. A review of the horticulture program at the university had made it clear that new opportunities lay in ornamentals and turf—not in tree-fruit production. Turf sales in Arkansas topped $250 million annually; tree fruit earned $2.5 million. Or put another way: Wal-Mart, the biggest corporation in Arkansas, had recently announced that it planned to open one thousand nursery centers in the next five years and it needed managers. Once, Rom said, he had approached a newly hired university executive and spoken excitedly about his expectations for fruit breeding. The administrator smiled broadly at the young horticulturist and explained that Arkansas chicken receipts earned more than the entire U.S. apple industry combined.

"We had to get going," Rom continued, "so I called up Mitch and told him, 'You won't believe this, but I've been told I can't plant 'em, that I might as well throw them in the trash.' Well, that was like cutting my heart out as an apple breeder."

"So I said, sure," Lynd recalled. "I'll take 'em. And why? I feel desperate to find a new apple that's a better quality than Rome Beauty, but we want one that has the reliability of the Rome. Because they bloom late, Romes just bear every year. But the time has come that Rome's place in the market is being seriously challenged."

Promising as those exotic Kazakh apples may appear, there could also be building blocks to future apples growing along the roadside just a mile or two from Mitch Lynd's house. We spotted two of them as we were driving out to his north orchard to look at some Jonathan. Gnarled old seedlings, their limbs twisting out of control, these apple trees were full of fruit. They had clearly beaten the frost, just as the Kazakh seedlings' parents had in 1993. There had been no research project launched to study these wildings, and a busy grower like Lynd had no time to undertake the collection, sorting, and initial nursery work. Roadside seedlings sprout wherever people toss apple cores out of car windows, but seedlings there in Licking County, Ohio, carry a special heritage. This was prime Johnny Appleseed country.

After leaving New York, and then northern Pennsylvania, John Chapman came to central Ohio in 1801. He planted half a dozen nurseries—all of them seedlings. It wasn't that Chapman didn't know about grafting. He wanted to do something else. He saw in the frontier of the American wilderness something akin to what the great American transcendentalists, Thoreau and Emerson, would later write about at mid-century. This New World in the Western Hemisphere, as the Puritans had seen it, held within it the prospect of universal redemption, a second chance for all mankind. Unlike the Puritans, the nineteenth-century utopians—call them transcendentalists, deists, pantheists, even gnostics—saw God in the vast expanse of nature, in seeds, plants, and the creatures of the forest floor. They looked to science, in particular agricultural science, as the method for revealing divine handiwork; therefore, they inevitably believed that the more they understood the natural world, the deeper their spiritual knowledge would be. For Chapman, man played a critical role in the expanding American prospect. He was not merely to see the natural world and note its characteristics; he was to engage with nature and develop it. He was to uncover the divine magic of things, bring them forth, make them manifest for all to see and use—in short, to recover the limitless natural diversity of the original Paradise in this new Eden known as the American wilderness.

Rough, self-reliant, usually barefoot, often dressed in nothing more than a potato sack cut with holes for his head and arms, this strange New England transplant who came to be known as Johnny Appleseed planted seeds precisely because he believed they would, in time, reveal the majestic diversity locked within the ruddy apple. Hundreds of new apples, never seen in Europe, would grow up across the American frontier —just as they had in the medieval groves that seem to have produced the seedling cider orchards of Spain and France. And, though John Chapman knew nothing of central Asia, these "wild" nurseries might in a generation or two grow into a sort of New World apple laboratory, not as diverse as the mountain forests of the Tian Shan range but richer in its diversity than anything else Americans and Europeans yet knew.

John Chapman was an eccentric American original, but he was not a total loner in his odd notions. He became, by the mid-1830s, America's most famous Swedenborgian. Emanuel Swedenborg had been one of the towering figures of eighteenth-century intellect, a man of science and a mystic whose followers have been as diverse as Ralph Waldo Emerson, William Blake, Jorge Luis Borges, and Nobel laureate Czeslaw Milosz. Swedenborgian metaphysics finds divine spiritual relations throughout the natural world. Each plant, creature, and object in nature corresponds to particular spiritual truths; interactions among elements of the physical world reveal ever more intense levels of spiritual insight.

For John Chapman, the scattering of apple seeds seems to

have served a dual purpose. On the one hand, his nurseries gave him access to the pioneer communities of the moving frontier. By propagating trees on these empty new farmsteads, he found himself well placed to proselytize for the cause of Swedenborgian faith. Simultaneously, his nurseries stood as laboratories of divine manifestation. Literally hundreds of new apple varieties appeared across the nineteenth-century American landscape, varieties that two generations earlier had not existed. They presented a utopian prospect no longer imaginable on the enclosed, overpopulated farms of Europe and England. Certainly, new varieties of apple, pear, plum, and other fruits continued to be found by the English and the Europeans, but by the end of the eighteenth century horticultural scientists were more concerned with identifying, cataloguing, and evaluating the apples that already existed. The seeming boundlessness of the American landscape only encouraged farmers to create new fruit trees alongside the grafts of established varieties. America, the utopians were convinced, uniquely contained the seeds of actual Paradise if only they could be released into her fertile, redemptive soil.

The redemptive apple had deep roots in the American experience. When John Winthrop and his Puritan sojourners landed on the shores of New England, they, of course, brought seeds and cuttings from their best apples. Apples were, after all, England's favored fruit. More, though, apples had become the culinary badge of Protestantism. Just as cider making and cider drinking had grown into a patriotic emblem for the English in their wars with the French, so the Protestants embraced the apple as spiritually superior to the grape, arguing that wine was endemic to the corrupt cultures dominated by Roman Catholicism. Apples, on the other hand, were a temperate fruit, born of a modest tree. They bore reliably (more reliably in the English climate, where spring bloom freezes were rare), and they required modest attention from the farmer. English Protestants even wrote treatises on the divine importance of orchards, including one titled *The Spiritual Use of an Orchard*, which argued that bitter, wild crab apples were the evil result of man's fall from grace in the Garden. Others attributed spiritual superiority to the apple because they believed, falsely, that it created fruit asexually, a vegetative monument to the presumed pristine and sexually undivided nature of Paradise.

Particular theological tics aside, the Puritans and their amalgamated Protestant cousins saw the apple as integral to their diverse visions of the New World. A small orchard of apples and pears was as vital to the new farmsteads as the annual sowing of corn—so much so that by the time the Northwest Territories were opened settlers applying for land grants were specifically required to "set out at least fifty apple or pear trees." Farmers who only tended cattle or planted row crops might easily abandon their land after harvest and join the westward march toward new speculative frontiers. The creation of settlements, however, required stability, and orchards were a demonstrable mark of stability since it took at least a decade for them to provide good supplies of fruit. Practically, apples were the only fruit that would last late into the winter, and, just as important, they provided a cheap, low-alcohol beverage for the farmer, his family, and his field hands. Even more than in England, cider became an integral part of early American social life, and by the outbreak of the Revolution, one of every ten New England farms was estimated to operate its own cider mill. First in New England, then in New Jersey and Virginia, cider became a sort of barter currency, offered in casks as payment for everything from shoes to a child's education. New Jersey won fame throughout the colonies for both its ciders and its distilled apple brandies, and New Jersey remains home to the country's largest commercial apple-brandy maker. Virginia's great plantations were never without cider, notably the Carter Grove plantation outside Williamsburg (where a stepchild version of the famous Hewes crab still exists), Washington's Mount Vernon, and Jefferson's Monticello. George Washington planted thousands of seedlings, tinkered with grafting and espaliered training techniques, and was a major cider maker. Thomas Jefferson's epicurean taste led him on a perennial search for fine, elegant ciders and wines. The wines were less successful, but his ciders seem to have won high praise.

Jefferson planted two orchards at Monticello, one below the back lawn for fresh apples and one just below the front drive exclusively for cider fruit. Esopus Spitzenburg, from upstate New York, was his favorite fresh apple, though he had a terrible time getting it to grow in the hot, humid Virginia climate. There was a single cider apple, however, that sent Jefferson into rhapsodies of delight. "They are called the 'Taliaferro' apple, being from a seedling tree, discovered by a gentleman of that name near Williamsburg," Jefferson wrote his granddaughter, with an accompanying package of scion wood, "and yield unquestionably the finest cyder we have ever known, and more like wine than any liquor I have ever tasted which was not wine." The Taliaferro (pronounced tolliver) was a small, late-summer apple. Its skin bore creamy red and white streaks, and red veins permeated the flesh—or so later accounts de-scribed it. Tom Burford, the antique-apple specialist whose an-cestors were growing apples at the same time as Jefferson, has worked with Monticello's curators, helping them rebuild Jef-ferson's orchards. Archaeological drillings have enabled them to locate the exact spots where each tree was planted, and Jefferson's detailed garden books identify the varieties. The Taliaferro, alas, disappeared.

Burford, who has won wide recognition as a careful apple sleuth, has been on the track of the Taliaferro for years. Ninety-six of them were planted at Monticello, but the last formal mention of them was in an 1817 edition of William Coxe's treatise on North American fruit varieties. In New Jer-sey, the famed Harrison cider apple also disappeared from for-mal records late in the nineteenth century. Burford told me when I first met him in 1995 that he thought he might have found both the missing apples. He explained how a doctor friend from Washington had been vacationing deep in the western hills of Virginia, in Highland County, where the thick mountain accent of the local people seemed to her almost Scot-tish in its lilt.

"The trees are on a hillside farm that belongs to a man named Colaw—Conley Colaw," Tom told me. Conley Colaw had mailed two of the little apples to Burford late in the sum-mer of 1994, and Burford grew very excited. "I think this is it," he said to more than one visitor, his voice a stage whisper, "but we can't say this for sure yet." All caution, with an air of intrigue, he counseled against any rush to judgment that these were truly the famed lost apples of Monticello, the apples that would ripen into as fine a "cyder" as the world had ever tasted. To make a confirmation, he would have to gather enough of the apples to press and ferment—a bushel or two at least. The sugar content would need to be measured, the color of the must judged, the full body and bouquet of the "cyder" assessed. The following year, Colaw's trees bore only a handful of apples, and somehow in 1996 Burford failed to get any of the fruit for a test. Colaw had given him some scion woodcuttings that will eventually produce their own fruit.

I wanted to see these mysterious trees for myself, however, and, with any luck, I hoped to taste the exotic apples. Late in October, on the way back to New York from Kentucky, I drove up into the remote hills of Highland County. The man behind the counter at the country store in the crossroads village of Bluegrass paused a second before giving me final directions to Conley Colaw's place. Conley, the storekeeper's facial muscles told me, was a special individual in the community's life. The first drive I turned onto, a rutted clay path more suited to tractors than cars, took me to an abandoned rock-and-frame house, where I found a barking hound chained to a post. That house, it turned out, had belonged to Conley's late brother. Conley's was the new, single-story house one driveway and two fields away. He was sitting in his pickup at the cattle gate when I backtracked to the correct entrance. He'd noticed the Kentucky plates on my car, watched me take the wrong turn, and figured I'd be back.

We stood beside his truck, a wet, gray sky close to the earth, and performed what a Kentucky friend of mine calls the pony-pawing routine, scratching the toes of our shoes on the ground as we circle around the subject at hand. Moving too quickly to the point breeds suspicion. I elaborated on what I'd told him by phone of my conversations with Tom Burford about the missing apples of Monticello. He talked about the farm, about the sheep I could see farther up the rounded, rolling hill, how the price of wool had fallen so low there was no money in shearing anymore, about the maple sugar shack he kept down near the road, and how much he could make selling syrup in the spring maple sugar festival.

"Guess you'd like to look at the apple trees, yeah?" he said after a while, and we set off toward a shallow ravine above which a grove of tall, twisted fruit trees stood. They seemed not to have been pruned for half a century. Their limbs shot up toward the clouds, thick and twisted like knotted strands of hair.

"That's a Laurie, that is, yeah," he told me, pointing to a smallish tree with red, Winesap-like apples that had begun to drop. All the fruit bore the stains of sooty blotch, scab cracks, and worm stings.

"And that's a Grimes Golden, yeah." An odd, staccato rhythm shaped his talk, every few words punctuated by a confirming "yeah." I supposed he was nearly seventy, though a handsome, sharp-featured seventy beneath the sun-weathered skin and gray stubble. In fact, he was only in his mid-fifties.

"That'n right over there, yeah, that's the one," he said as we walked to the bottom of the grove. "The tal-i-fera. That's what they think it is. Yes they do. I don't know. It's the one I sent to Monticello."

No apples hung on the tree. They ripen in August, and he'd picked them two months earlier.

"Red with white stripes," he said. "Then they gets real dark red with white spots. It gets red ripe, gets red inside, yes it does."

I was prepared to leave appreciative but disappointed. A bare autumn apple tree doesn't tell you much. When I had telephoned him from Kentucky, he hadn't told me these were summer apples. We were walking back up the open field to his house. Relaxed, he was telling me a little of his family history—how his ancestors had gone early to California and how his grandmother had come back, carrying a sapling that had grown into the sturdy Monterey cedar that still stands beside a second abandoned house twenty yards from the apple grove. None of the family lore included any details about the apples he'd learned to call Taliaferro. Just about to my car, he asked, "Want to taste some of it?"

Inside the kitchen, where the walls were lined by cupboards and cabinets he crafted from walnut, butternut, and cherry, he pulled a gallon jar from the refrigerator. The liquid inside shone a rusty rose. Sediment covered the bottom. A few apple bits and fragments of leaf floated on the top. He dipped a kitchen ladle into the still juice, now two months old and nearly clear. Without any question, it was the richest, fullest-bodied farm cider I'd ever tasted, nearly dry. It had a finish almost like fine sherry but full of apples. Single-variety ciders are usually dull and flat; this one, whether or not it was Mr. Jefferson's pride, was the best I'd ever swallowed.

But for the fateful turns of religion and immigration, ciders made of apples like these, ciders that were more common drinks in the countryside than beer and water, might have won worldwide acclaim. The Taliaferro, the Harrison, the Hewes or Virginia crab, and even the Winesap were all originally cider apples that sprang from seedlings. As they had been in England, Normandy, and earlier in Asturian Spain, seedlings, their fruit usually too bitter to eat raw, were the mothers of cider and of most new apple discoveries. So long as the seedling plantings lasted—and John Chapman's Ohio plantings were the last commercial seedling nurseries of any size—cider remained a vigorous part of farming and the rural economy. Or, perhaps better, so long as cider remained a country beverage, apple seedlings continued to be of value. Three implacable forces, however, all but destroyed American cider in little more than three generations: German immigration, the Christian temperance movement, and urban industrialization.

Cider seems to have reached its American apogee in the Presidential election of 1840, when William Henry Harrison, from Ohio, pursued his "Log Cabin and Hard Cider" campaign on the Whig ticket. Harrison's campaign presaged the gaudy hoopla mixed with nostalgia that we think of as a twentieth-century phenomenon: campaign hats, flyers, ad placards, staged rallies, and a single, unifying insignia—a log cabin with a wooden barrel marked HARD CIDER at the doorstep. Everywhere the campaign's advance men went, cider barrels followed. By 1840 cider probably had already begun to slip behind beer in popular taste, but it provided the former war hero with a valuable image. The Whigs successfully painted portraits of the incumbent, Martin Van Buren, as an extravagant "gold spooned" aristocrat while casting General Harrison as a simple country yeoman who sipped the drink favored by the Founding Fathers. In an age when German and Irish immigrants were already throwing a new—and to many, a disturbing—cast on urban life, filling jobs in recently built factories, crowding together in worker dormitories, swilling beer and whiskey in big-city taverns, the cider barrel stood for an older, cleaner, purer America. Perversely, Harrison won, but cider lost.

This longing for an earlier, purer America coincided with the birth of the temperance movement. All through the Colonial era and the eighteenth century, the Puritans and other Protestants had issued stern sermons about the sins of drunkenness and public disorder. Still they drank. Cider and low-grade beer were household staples, as was corn whiskey throughout the South. But hardly anyone supposed drink to be innately evil or capable of creating disease and dependency. Drunks were simply people who had failed to exercise self-control. As America set out on its long, steady march away from agriculture and toward the cities, as the process of urbanization was launched, much of the old English, Protestant stock began to feel alarmed. Corruptions of the Old World were crossing the Atlantic. Baltimore, Philadelphia, Boston, and New York grew increasingly crime infested, as later Pittsburgh, Cleveland, and Chicago would. Indeed, by the end of the century, guns and gangs had turned the big Eastern cities as violent as they would become again in the late twentieth century.

Early reformers, fueled with anti-immigrant and anti-Catholic fervor, found the answer in demon liquor. "Temperance supporters," historian Harry Levine has written, "believed they had located, in liquor, the source of most social problems. The Temperance Movement, it should be remembered, was the largest mass movement in 19th-century America. And it was an eminently mainstream middle-class affair. The Temperance Movement appealed to so many people, in part because it had become a 'fact of life' that one could lose control of one's behavior. Even the use of the word 'temperance' for a total abstinence movement is understandable when we realize that the chief concern of temperance advocates, and of the middle class in general, was self-restraint. Liquor was evil, a demon, because its short- and long-run effect was to prevent drinkers from living moderate, restrained, temperate lives."

By the eve of the Civil War, temperance and abolition had become linked as a progressive reform movement. Yet for all the temperance movement's successes, it was only cider, the most benign, most traditional, and most "Protestant" beverage, that genuinely suffered. David Williams, an amateur historian and cider enthusiast, has argued that cider fell from favor precisely because it had been the household staple of the rural WASP families and it was, therefore, the one drink they could effectively purge. Beer drinkers were disproportionately city people, disproportionately Roman Catholic—and priests were nearly never temperance campaigners. German-speaking immigrants outstripped Anglo newcomers to the degree that now there are more Americans of German ancestry than of English. Because they retained their language and their drinking tastes in their own neighborhoods, they were effectively insulated from the Protestant reform agitators. It might be added as well that many mill owners surely understood the riskiness of eliminating one of the few cheap pleasures their workers possessed.

Finally, aside from the social forces aligned against it, cider suffered in nineteenth-century America as it would in twentieth-century France. Cider had been a drink of the rustic life, fermented at home or no more than a few miles away, and by the 1890s the population had already begun to depart the countryside. In the days before cold storage, apples had to be harvested and milled quickly. They were too bulky and perishable to take to town in great quantity, a problem of far greater import in America than in England because of the great distances and poor roads between cities in the United States.

Unfiltered, unpasteurized ciders never traveled well, and they still don't, whereas grain for beer was easily hauled and stored in city warehouses, where it was held until the brewery needed it. Beer, simply, was a light drink better designed for the ever more urban population, and that population either had no memory of cider or, if it did, was often eager to forget the crude beverage of a crude past.

As steadily as cider declined, however, apples continued to flourish. If the temperance reformers had painted the evil cider apple as corrupter of frail humans, the hygienic movement came to its rescue by the century's end. Healthy, red-cheeked lunch-box apples, now conveniently shipped aboard railroad refrigerator cars and stored through the winter in icehouses, were resurrected as icons of wholesome living. Even smelly, barefoot Johnny Appleseed, that follower of a foreign mystic with an exotic name, bounced back as a sort of Midwestern Yankee Doodle suitable for children's fables. From 1885 until 1950, no fewer than three hundred books, plays, and operas about the old Swedenborgian were published. Apples as never before became the hyper-American fruit. Boosted by the damming of the Columbia River, huge new orchards covered the banks of the Wenatchee and Yakima Rivers. Irrigation gave the apples perfect size and shape. The clear desert sun and cool nights painted them with deep, luminescent hues that consumers had never seen, apples as clear and shiny as giant, polished rubies. At last the fruit of Paradise had become a full partner in the age of science and modernism, along with the automobile, the assembly line, and mass-market advertising. The Protestant apple orchard, that private garden of grafted favorites and wild seedlings, so integral to a family's food, drink, and geographic stability, had not died but instead was reborn as a living food factory that required engineers, chemists, geneticists, and marketing brokers for its survival.

Mitch Lynd stood over his collection of potted Kazakh apple seedlings full of pride and puzzlement. He had gone to Ohio State in the 1960s and studied economics. "We thought we knew about horticulture, but we thought we needed to know a lot more about how to survive commercially," he told me. The first in his family to graduate from college, Lynd had spent three years working for Kroger, the giant Midwestern supermarket chain that had been buying the family's apples and peaches. He decided to return to the farm early, three years out of college, but only if his father would do what the New York and Washington growers were doing: triple the size of the orchard, build sophisticated cold-storage units, and work the national markets on an almost year-round basis—in short, enter that elite club of perhaps two dozen packer-growers who defined the North American apple industry.

Now, thirty years later, the Kazakh seedlings presented a dilemma and an opportunity. Apple growers, large and small, stood at the threshold of a new era. In his great-grandfather's time, in the 1880s, this very farm produced more than a hundred apple varieties, every one of them a known nursery cultivar. Seedling plantings were a distant bit of nostalgia left to the Johnny Appleseed biographers. The dawn of scientific apple breeding was just opening, developed by state land-grant universities and largely funded by the federal government. For a full century, that system had brought prosperity to huge orchards like the Lynd farm, and it turned New York and Washington into the world's leading apple producers. But by the 1980s two things had changed: public finance of agricultural research fell from fashion, and consumers began again to demand more than cosmetics in their fruit. They wanted apples that tasted good. They were ready to drive directly to farm markets to buy them and often to pick the fruit themselves. What they didn't want were Rome Beauties and Red Delicious. Mitch Lynd saw the shift coming. In the 1980s he put in about twenty acres of pick-your-own apples, and by the mid-1990s those twenty acres were providing half the farm's income. Still, he needed new apples, apples that were as reliable as Rome Beauties but that tasted good when you sank your teeth into them.

"There are so many really *good* apples out there now. Suncrisp. Fuji. Jonagold. They're a whole lot better than Rome and Red Delicious, but a cold spring like this year's kills them," he said. That's the reason he snapped up the homeless Kazakh seedlings. Perspicacious curiosity had kept him current with the Cornell expeditions to Kazakhstan; a sharp eye for changing market conditions led him to take the seedlings in. "I just think the answer may be this Kazakh material. The basic building blocks of the genetic map to new cold-hardy apples may lie right here."

But who would do the breeding? Ohio State and Michigan State had no serious breeding programs. Purdue still maintained a modest one. Arkansas—well, the answer there was clear. Cornell ran the only major breeding program, but its interest was foremost in creating apples that would serve New York and Northeastern growers. A new program in Washington would focus on fruit that flourished in the climate of the Pacific Northwest. What the Lynds needed was what we needed in Kentucky, which is what small and medium-sized apple growers need all the way from the Mississippi to the Chesapeake Bay: apples that can withstand the climate and the phalanx of pests that flourish in that climate, and further, apples that possess a wide array of flavors, colors, crispness, and ripening dates for an increasingly picky retail market. We are looking for orchards that grow modern-quality fruit, that will permit us to reduce dependency on chemical controls, and that contain the diversity that farm orchards had in Johnny Appleseed's day.

Mitch and a nurseryman from Indiana named Ed Fackler hatched a plan. It was the sort of plan Thomas Jefferson or George Washington or John Chapman, or even the horticultural clubs of eighteenth-century England, would have loved—to a point. They and a handful of other growers and nurserymen would form the Midwest Apple Improvement Association. Working with apple specialists at Ohio State University, they would organize a cooperative breeding program to create new apples tailored to Midwest growing conditions. For one hundred dollars a year, any grower could become a founder and gain early access to the new varieties being created. Each of the founding growers would take a few dozen of the Kazakh or other known cold-hardy, disease-resistant seedlings and plant them in remote locations adjacent to selected high-quality cultivars like Fuji, Gala, or even such favored antiques as russets, Mothers, and Seek-No-Furthers. Then they would wait for nature to take its course, relying on bees and wind to cross-pollinate the trees, just as she had done

in the wild for millennia, thereby eliminating the tightly controlled and extremely expensive pinching and pollen painting
that goes on in greenhouse breeding. University specialists
would help the growers screen for resistance to scab, fire blight,
powdery mildew, and other diseases, at the end of which each
orchard would plant as many as one thousand of the crossed
seedlings. If twenty orchards joined the program, the Association would have twenty thousand different seedlings growing
across the Midwest, from which two or three, perhaps half a
dozen, good apple varieties might emerge. In effect, they
would be replicating the old, prolific "wild" cider orchards,
which were the unorganized laboratories, or gene pools, that
gave birth to many of the hundreds of eighteenth- and nineteenth-century North American apples. The big difference
now, however, would be law and property rights. These new
Midwestern specialty apples could be grown only by members
of the Midwest Apple Improvement Association, which would
trademark the names of the apples and control who could sell
them on the open market.

Trademarking an apple, as though it were a soft drink or a
cereal brand, would surely have shocked the eighteenth- and
nineteenth-century agricultural utopians. Discover God's wonders, they counseled, then spread them far and wide as evidence of divine munificence. Privatization of public research
and the global evolution of intellectual property law have in
the last twenty years changed the old precepts. For most of the
last century, fruit breeders have created apples like Cortland,
Jonagold, Gala, or Fuji, then patented them. Nurseries charged
a royalty of fifty cents or a dollar on each tree sold to repay
the breeders for their development costs. Agricultural patents

and copyrights, however, last only seventeen years, after which rights to the cultivar are free. Since it takes eight or ten years to get the trees into production and develop a market for the new varieties, patent holders really made money for only a decade. Trademarks, on the other hand, last forever, so long as they are renewed regularly.

New Zealanders, who developed the Braeburn, were the first to begin trademarking the names of their apples. One of their newest is called Pacific Rose. "They said never again were they going to do the research to make the apple, market it, develop the demand for it, and then give it away after ten years of patent royalties. Now, everything's becoming proprietary," Lynd explained. "We could grow Pacific Rose if we got the wood. We just could never call it that. The effect is that, no matter how popular a new apple becomes, I'll never be able to grow it. If we don't get off our butts and develop our own apples, we may as well roll over and play dead. We'll be out of it."

Under the new playing rules, nature and property rights are reconfiguring both local and global access to food, for what is true of apples will surely soon be true of the world's major edible commodities. What seems likely is that a new sort of food tribalism will emerge. Trademarked Washington State apples will certainly be sold by Kroger's in Columbus and D'Agostino's in New York. And high-flavor Midwestern or New England apples, suited specifically to the soils and climates of their origin, might be sold in Santa Monica and Seattle. Trademark groups would likely try to hold prices high by restricting how many of their apples could be grown or sold to brokers and grocery chains. Just possibly, a new kind of competition based on intensity of flavors and regional specialties could return some of the rich variety of apples that nearly disappeared in the 1950s, 1960s, and 1970s as western Red and Golden Delicious swamped the market. Such trademarking rules might also grant salvation to small orchards selling prize fruit directly to consumers. Or, alternatively, regional and national trademarking might create global cartels that effectively eliminate any independent growers from selling their fruit to the supermarket chains and wholesale markets. Trademark cartels, as buyers and sellers of computers have learned, offer no guarantee that the best products will be rewarded or will even survive in the emerging global marketplace.

Pomona, the Romans believed, was the country nymph-goddess who gave fruit to the earth. She tended her orchards, pruned and shaped them, and taught man how to graft one variety onto another. Vertumnus, the harvest god, longed for her love and worked diligently in the fields to win it, but always Pomona spurned him. She was determined never to wed, until one day an old man came into her orchard begging to know how one so beautiful as she remained alone. Pomona teased the old man, but he argued and parried with her until at last she swore that of all her suitors she found only one, Vertumnus, worthy of her love. The old man shed his wrinkled body and revealed himself to be Vertumnus. Realizing how cleverly she had been caught, Pomona released herself to Vertumnus's embrace, married him, and together they brought the sweet fruits of her orchard to harvest.

No matter how much the crass and tacky laws of the marketplace threaten to transform Pomona's fields into assembly lines for bland commodities, the temptations of the blushing goddess seem somehow to survive. Mitch Lynd, who knows more than a little about both taste and commodities, retains his optimism. We were talking about his notions for organizing an apple improvement association and the prospects for sustaining fine fruit in the global economy. "Here's the key," he said. "People aren't coming out here to pick apples because they need food. They're coming out here to see and touch something that feels real. They don't care whether it looks all shiny and perfect. The more technology and mega-chain markets take over, the more human beings need to have some sort of counterbalance, some way to stay in touch with actual living systems. People have a deep-seated need to be with, even be surrounded by, a living system. Well, that's what an orchard is."

For all but a few of us, the food we eat comes from unseen, untouchable origins. Corn and wheat sprout on gargantuan fields too immense to imagine in personal terms. Broccoli, lettuce, cauliflower grow from the irrigated floors of blistering deserts. Stockyards, most of us would rather not think about. Apples, peaches, and plums mostly ripen on immense factory fields to be sorted by computerized grading machines, preserved by digitalized atmospheric controls, marketed on the Internet by fast-talking brokers in Stockton and Wenatchee juggling container shipments to Taiwan, Singapore, and Riyadh. Still, even they succumb finally to the sweet fragrance and seductive juices of a single piece of fruit, which, when grasped between thumb and forefinger and drawn to the mouth, reminds us that the art contained within that solitary apple is more powerful (and surely more subversive) than the greatest commodity cartels. Sweet-sour juice trickles across the surface of the tongue, bathes the eater's mouth, and the apple becomes its own temptation, returning us to our remembered, or imagined, childhoods, to October mornings just after frost has melted into dying grass and we leaned against the scaling bark of a bent and heavy limb and drifted into a morning reverie on the prospect of that single apple, hanging alone on its stem, ready to drop this very day, this hour, and thereby conclude the orchard's work. Finally, it is not merely the apple we enjoy in its perfection but the transporting journey on which it takes us, to a Paradisiacal garden, where the grubby world of tasteless commodities was never known and the legions of devouring pests are all vanquished.

That one apple carries us into the garden of our imaginations, where, the English horticulturist William Lawson wrote nearly four hundred years ago, "all delight in orchards, for whereas every other pleasure commonly filles some one of our sences, and that onely, with delight, this makes all our sences swim in pleasure . . . What can your eye desire to see, your eares to heare, your mouth to taste, or your nose to smell, that is not to be had in an orchard . . . ?"

APPENDIX I

Twenty (or so) Prize Apples

Every year at our orchard, customers ask us, "What's a good eating apple?" and "What's a good cooking apple?" Our answer is to ask a few more questions: Do you like to eat tart apples or sweet ones? Hard and crisp, or a little more "mellow"? How do you plan to cook them? Should they stand up firm for baking? Do you want to make a sauce or an apple butter? If you're making a pie, do you want whole pieces of apple when it's finished, or do you want it more like a pudding or a sauce pie? With all those questions in mind, the following is a selection of wonderful apples for all purposes, with notes on their origins, uses, and characteristics. Some are quite new, a few are fairly old, but all of them can be found from commercial nurseries.

· · ·

Belle de Boskoop (France, Belgium, Netherlands): Large, slightly flat, russety red over pale green, sometimes with orange-red stripes. Pale yellow flesh beneath tough skin; sharp, tart flavor; holds shape in baking. Found 1856 by K.J.W. Ottolander at Boskoop, Netherlands, on a sport limb of Reinette de Montfort.

Black Twig (U.S.): Medium to large, dark red over green, often with rough finish; picked in late October. Hard, crisp, yellowish flesh, slightly more tart than sweet; holds shape in cooking; keeps well through the winter. Morgan and Richards attribute it to a Winesap seedling near Rhea Mills, Arkansas, with commercial introduction in 1868; Burford identifies its origin as an 1830 seedling from a farm near Fayetteville, Tennessee. Popular around turn of the century in the South, especially central Virginia, but dropped by 1930s for poor yields.

Bramley's Seedling (Britain, Ireland): Large, pale green with occasional reddish-orange stripes. Creamy, coarse, acidic flesh; used as the definitive English "cooker." Grown first by Mary Ann Brailsford in a garden in Southwell, Nottinghamshire, England, 1809–1813; named for later cottage owner, a local butcher named Bramley; first exhibited in 1876 and commercially grown in the 1890s and after.

Calville Blanc d'Hiver (France): Medium-sized to large, pale, yellowish green with deep, sometimes almost star-fruit-like pentagonal ridges; picked in November for late keeping. Very tart at picking, mellowing to rich, complex taste; holds shape; traditional choice for tarts. First recorded 1598 as Blanche de Zurich, 1628 as Calville Blanc, rare now but greatly beloved by chefs; very high in vitamin C.

Cox's Orange Pippin (Britain, Belgium, France): Medium-sized, golden-orange blush and occasional stripes over yellow base, often with a light dusting of russet. Rich, creamy semi-crisp flesh; sweet, intensely aromatic with spicy, almost nutty overtone. Believed a seedling of Ribston Pippin grown by Richard Cox at Slough, Buckinghamshire, England, about 1825; commercially introduced about 1850. Numerous variations; moderately high in vitamin C.

Esopus Spitzenburg (U.S.): Medium-sized, bright-red blush over light yellow with faint ridging. Hard, crisp yellow flesh; sweet and fruity with tingling tartness; keeps well in storage; holds shape in cooking, excellent in pies, and possible inspiration for Waldorf salad. Planted in Ulster County, New York, prior to 1790; favorite of Thomas Jefferson; widely planted in nineteenth-century United States.

Fuji (worldwide): Creamy yellow overlaid with orange-red blush, medium-sized to large, with tough skin and crunchy-crisp, honey-sweet yellow flesh. Slow ripening from September to November requires multiple pickings. Commands high prices but difficult to grow and prune. Developed in 1939 by Horticultural Research Station at Morioka, Japan, from Ralls × Delicious; named and released in 1962.

Golden Delicious (worldwide): Medium-sized to large, bright yellow, sometimes lightly russeted in humid climates. Sweet, crisp, exceptionally juicy and aromatic if grown in native mid-Appalachian zones and allowed to ripen fully on tree but often mealy and bland in Western, Midwestern, and European zones. Yellow flesh holds shape when cooked. Found as chance seedling, possibly of Grimes Golden, on farm in Clay County, West Virginia, near the Ohio River, in 1890; introduced in 1914.

Goldrush (U.S.): Medium-sized, bright, shiny yellow with dark-pink flush. Pale, very hard at late-October picking; strong tart flavor that reaches full sweet-tart balance one to two months after picking and holds crispness in storage for six months. New, disease-resistant cross of Golden Delicious and unreleased "Co-op 17" in early 1990s from Purdue, Rutgers, and University of Illinois cooperative breeding program.

Honeycrisp (U.S.): Large, scarlet-orange splash over pale yellow. Mellow-sweet and fragrant; crisp and full of juice; ready in early autumn but hangs on tree for up to five weeks and holds crispness until midwinter. Holds shape as baker. Developed by University of Minnesota from Macoun × Honeygold as a winter-hardy variety; released in 1991.

Jonagold (U.S., Belgium, France): Large, orange-red blush covering up to 80 percent over greenish-yellow ground. Crisp, yellow flesh combines nutty Golden sweetness with tart Jonathan fragrance and holds shape for frying or pies. Crossed 1943 at Cornell Agricultural Experiment Station, Geneva, New York, and released 1968. Numerous sports provide highly varied color and finish; marketed under, notably, Rubinstar and Jonagored.

Jonathan (mostly U.S.): Medium-sized, bright to deep red over green. Deep, complex, sweet-tart and strong, wine-like fragrance. White flesh holds its shape well in stewing or baking. Long regarded as an excellent all-purpose apple but a poor keeper even in cold storage. Appeared on farm in Woodstock, New York, possibly a seedling of Esopus Spitzenburg; described first in 1826 and named for Jonathan Hasbrouck, who brought it to the attention of the Albany Horticulture Society and thence to Massachusetts Horticulture Society.

King David (U.S., eastern Europe): Medium-sized to small; deep, inky red with occasional yellow undercolor. Rich, complex balance of high acid and intense, aromatic sweetness with a slight, astringent edge. Crisp, yellow flesh full of juice; holds shape for baking or frying. One of the few excellent single-variety ciders. Found 1893 as a seedling in a Washington County, Arkansas, hedgerow, a possible natural cross of Arkansas Black and Jonathan or Winesap and Jonathan.

McIntosh (U.S., Canada, northern Europe): Medium-sized, bright red to deep purple over green, with a dusty bluish bloom. Crisp but not hard white flesh with berry-like aroma and high juiciness. Short storage life, cooks quickly to a sauce. Discovered by John McIntosh in 1796 in Dundas County, Ontario, Canada; possibly a seedling of Fameuse (or Snow Apple); distributed by son Allan around 1870. Parent to Cortland and Empire.

Mother (U.S.): Large, red flushed and striped over green, somewhat lopsided. Crisp but not hard, ripening to soft and juicy. Extremely popular nineteenth- and early-twentieth-century dessert apple, with a complex taste and aroma variously described as "balsamatic," like "pear drops," or spicy "vanilla." Appeared at Bolton in Worcester County, Massachusetts, and recorded 1844, with recent resurgence for specialty orchards.

Newtown Pippin (U.S.): Medium-sized to large, pale green ripening to yellow. Hard, crisp, and juicy; reaches its prime after one or two months' storage; noted for late keeping. Famed in the nineteenth century as the apple that launched American apple exports with private shipments to Queen Victoria; noted for its brisk, aromatic, rich pineapple flavor. Favored pie apple that mostly holds its shape. Found on estate of Gershom Moore, Newtown, Long Island, and well known by 1759; introduced in England by Benjamin Franklin. Now moderately available in Washington, Oregon, and California.

Russets (worldwide): Medium-sized, golden to bronze, grainy-textured skin often over greenish underskin. Dozens of russets proliferate in Britain, northern Europe, Canada, and formerly in the United States. Most share a pear-like crispness with moderate juice and keep well but gradually lose their snappy texture. St. Edmund's, dating from 1875 in England, bears a greenish skin beneath a fine grainy russet and, when allowed to ripen fully on the tree, was prized for its rich, creamy pear-like qualities. Roxbury, dating from the early 1600s in Massachusetts, is similar but has a coarser, almost crumbly flesh. Golden, dating from a century later in New England, is rounder and more thoroughly russeted than the earlier two. Razer Russet, found by Gene Razer in Eastern Kentucky, and Russet Beauty, identified by Indiana nurseryman Ed Fackler, are two modern russets that have indistinct origins in the 1960s, are larger, and bear a finer, pear-like finish and strong aromatic flavor.

Stayman Winesap (U.S., Middle East, Italy): Medium-sized to large, red or red striped over green, with a faint violet cast. This favorite of all the Winesaps has a bracing sweet-tart freshness but has fallen from commercial favor due to the apple's tendency to crack open after summer storms. The yellow flesh holds its shape slightly. Keeps through the winter without refrigeration but loses its crisp snap. Raised in 1866 by Dr. J. Stayman at Leavenworth, Kansas, from a Winesap seedling and introduced 1895 by Stark Nurseries, in Missouri.

Suncrisp (U.S.): Large, golden yellow with up to 50 percent orange blush. Crisper than Golden Delicious and most other yellow crosses; flavor improves with one to two months' storage and lasts well into winter with common storage. New variety that has won raves from blind tasting panels. Introduced by Rutgers University Experiment Station as a cross between Golden Delicious and Cox's Orange Pippin.

And Don't Forget the Crabs! Technically, a crab apple is any apple that falls below two inches in diameter. Several crab apples are used regularly to enhance pollination, largely because they have dense, long-lasting blooms and are heavy with pollen. Manchurian blooms early to mid-season, and Snowdrift blooms from mid- to late season; both bear white petals. Indian Summer blooms early and lasts well past mid-season; its flowers are full and pink. Other crabs, however, not only provide excellent pollination but also grow into handsome trees and bear genuinely delicious fruit that makes an excellent addition to cider. Hewes Crab, thought to be a seedling of Thomas Jefferson's beloved Virginia Crab, is sweet enough to eat despite its astringent skin, and Chestnut is a new, yellow crab from Minnesota with a bright scarlet blush that is wondrously delicious fresh.

Exact bloom and harvest dates of all these apples vary with latitude and local climate. Unless otherwise noted, most of the above will ripen between September 20 and October 20 in the Ohio Valley and central northern Virginia. For Tennessee, Arkansas, central California, and North Carolina, assume a week to ten days earlier. For Michigan, Pennsylvania, New Jersey, and the Hudson Valley, assume a week later, and for the Rochester–Finger Lakes area of New York and for New England, assume ten days to two weeks later on most varieties. It's always a good idea, however, to check with local agricultural extension horticultural specialists or with local commercial growers to establish accurate seasons.

APPENDIX II

Back-Yard Orchards

For nearly a thousand years a small orchard was a part of nearly every viable farm in northern Europe, and later in the northeastern United States. Apples being the hardiest fruits, the apple orchard would hold at least a dozen trees of several varieties—some for fresh eating, some for cooking, some for making cider. Most of those trees were grafted onto standard, or full-size, rootstocks that could grow to a height of thirty to forty feet. Even in the Middle Ages, however, a few small, or dwarfing, trees had been found, notably the Paradise, that seldom grew to more than ten or twelve feet and produced small bright-yellow apples. During the Renaissance, Paradise and other apples grafted onto Paradise rootstock were favored in elegant "pleasure gardens" of grand estates in France and England. The French particularly liked to make espaliered trees that could be mounted against garden walls or alongside decorative fences.

Today both the commercial and the back-yard orchard have a dizzying variety of choices available for apple tree size and shape, though the great, old-fashioned, sprawling standard tree has become rare. What follows is a guide to three broad categories: dwarf or bush trees; semidwarfs; and semi-standard.

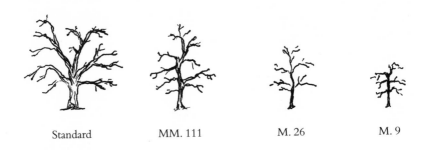

Standard   MM. 111   M. 26   M. 9

Dwarf Trees: The most common rootstock is M.9, named for the East Malling Research Station in Britain, which first developed and categorized the M series of rootstocks. A later collaboration with the John Innes Horticultural Institute at Merton produced the MM designation for a second rootstock series. The Budagovski, or Bud-9, is similarly sized and is resistant to some diseases that affect the M.9. Slightly smaller is the M.27, at about a quarter the size of a standard tree, and the G.65, developed by Cornell at Geneva, New York, as a more disease-resistant stock. M.26 and G.11 produce slightly larger trees, about 40 percent of standard size. *All the dwarfing rootstocks require either staking or trellising to prevent being blown down.* Selection of a particular tree depends on soil quality, drainage, the type of apple cultivar being grafted to it, and the overall orchard design desired.

Semidwarf Trees: M.7 has for thirty years been the standard semidwarf tree planted from upstate New York to the Pacific Coast. M.7s generally produce trees fifteen feet tall, or about half standard size, and they tend to be heavy bearers. G.30 from Cornell is a new arrival in the same category that appears to provide better disease resistance as well as earlier and heavier fruit cropping. Like the dwarfs, both should be staked—at least for the first several years after planting—but trellising is not required.

Semi-Standards: MM.106 rootstocks make trees about two-thirds standard size, eighteen to twenty feet tall, and MM.111s can grow to three-quarters full size. MM.106s have been heavily planted in Europe, especially in dry, well-drained soils, but they are extremely susceptible to collar rot, a fungal disease that kills tree tissue at ground level. MM.111s are probably the most care-free rootstocks; they require no staking, adapt to most pruning styles, tolerate wet soil better than most others, and withstand hot summer weather, all of which has made them a favorite of growers in the South and lower Midwest.

Note: Tree sizes and shapes vary with individual apple cultivars (e.g., Stayman Winesap and Braeburn are less vigorous and grow to smaller size on dwarfing stocks than Jonagold and Cortland) and with climate. An M.26 in Kentucky or North Carolina may put on as much growth in a season as an MM.106 in upstate New York because of the longer growing season in the southern latitudes. Therefore, tree spacing and training must be adjusted to local conditions.

Aside from rootstock considerations, home or garden orchardists will also need to choose the shape and design of the trees they want. Most northern and western commercial growers are planting dwarfs at five hundred to one thousand trees per acre, that is, three to six feet between each tree, and twelve to fifteen feet between rows, stabilized with individual stakes and/or wire trellises. Sometimes they pull every other tree to an alternative angle of forty-five degrees to produce V-shaped trellises that feed extra sunlight into the trees. Others, following a French system called the Vertical Ax (for axis), avoid almost all pruning and tie down the tree's lateral limbs to retard wood growth and get early fruiting. Standard and semi-standard trees traditionally have the central leader removed at three feet with three or four large, sturdy limbs left to form an "open vase," or bowl shape, that draws light into the center of the tree.

If your object is to harvest your own specialty apples, and to do it quickly, Vertical Ax is the method to choose, but bear in mind that it requires almost weekly attention during the first two or three summers to get the trees trained to a proper shape. If you want a pretty, lofty tree that invites climbing, then choose a larger stock and prune it to the open-vase style. If you want something in between the two, select a semidwarf, preserve the central leader, and prune the laterals at ninety degrees from it. Following are four possible choices: open vase, central leader, staked Vertical Ax, and espaliered.

Open Vase   Central Leader   Staked Vertical Ax   Espaliered

A Few Good Recipes
from Around the World

So far as we know, apples have been pleasing the human palate since about the end of the last Ice Age. Food historians frequently comment on the carbonized remains of small, crab-like apples, apparently roasted, found in the Swiss Alps. Nordic myth teases us with the image of the gods nibbling on apples to preserve their immortality, and the Egyptians, under Ramses II, certainly enjoyed apples and may well have fermented them into some sort of *sydra*. Of the great early Mediterranean and Persian culinary cultures, however, not one seems to have thought much about cooking apples. Even into the European Renaissance, apples were largely eaten fresh, either as a break between courses or as a sweet at the end of the meal. For the last several centuries, however, apples and fermented apple ci-der have claimed a vital place in the culinary repertoires of chefs all over the world. There are three broad categories of apple cookery: meats, soups, and savories; desserts; and cider-based dishes. Many of the basic dishes cross cultures and continents, while some are local and specific. The guide below surveys several of the most interesting apple dishes and aims to illustrate how they vary from land to land.

Meats, Soups, and Savories

Apples are combined with roasts, chops, and sausages all across northern and eastern Europe, Turkey, and South America. Apples show up almost everywhere with roast pork, largely because both are the standbys of peasants and the poor. Either tart or sweet apples may be used, depending on the desired effect. Tart acidic apples, cooked slowly in a pork stew, with the addition of cider vinegar as a marinade, help tenderize and break down muscle on tougher cuts. Sweeter apples, like York, Golden Delicious, Gala, and some of the russets, hold their shape—especially if added midway through the cooking time —and add both texture and color to the dish. Chop the apples *with their peeling intact* along with onions and herbs for a heartier dish.

Greeks, Turks, and Iranians make similar roasts and stews using lamb instead of pork. Moroccans often blend dried fruits or fresh apples into lamb stews, called *tagines*, that are first cooked over a direct flame and then finished in the oven. The lamb is cut into large cubes and coated with a seasoning paste of turmeric (or, more elegantly, strands of saffron), pureed onion, salt, pepper, ginger or cumin, ground red pepper flakes, and chopped parsley, then gently stir-fried in butter, covered with water, and simmered for an hour or so, and then two more coarsely chopped onions are added for texture. Once the meat is tender, it should be transferred to an open casserole and surrounded by lightly caramelized sliced apples—Jonathan or Jonagold would do well—and the rest of the liquids reduced to a sauce and poured over the lamb before putting it into a medium-hot oven for twenty minutes.

The secret is to balance the earthy, aromatic meat spices against the sturdy, sweet fruit, which has been sharpened with a little cinnamon, ginger, nutmeg, or allspice. Toasted almonds are frequently added just before serving to any sweet-savory *tagine*.

One of the most unusual stews with dried fruit and apples comes from Argentina and uses either beef or lamb, onion, green pepper, garlic, bouquet garni, tomatoes, serrano chilies, and beef broth, after which comes an improbable combination of white beans, a cubed sweet potato, a cubed white potato, two zucchinis, two ears of corn, a cup or so of dried apricots, and two cups of diced apples—served in either a casserole or a hollow pumpkin.

Sausages are widely combined with apples as well. Fried apples, biscuits, and link or patty sausages are a classic Saturday night supper throughout the American South. Simply slice and core (*but don't peel!*) the apples, toss them with sugar (or brown sugar if you would like a heavier dish), and fry in butter until the apples become crispy but soft on the inside. I've had so-called nouvelle-style fried apples, usually Granny Smiths, in stylish West Coast restaurants, where the apples remained crisp;

avoid that technique unless you also like your biscuits made with millet and sunflower oil. One of the most famous German peasant dishes, called *Himmel und Erde*, uses sliced apples and potatoes stewed together with a little salt, pepper, and sugar until tender, then combined with chunks of bacon or blood sausage and sautéed onion. A Scandinavian version, called *Äppel-Fläsk*, uses Canadian bacon and sliced apple rings cooked in a skillet on top of sautéeing onion slices.

Game is often roasted or braised with apples as well. Two tart apples or an apple and an onion are often used to stuff a duck or goose cavity in Norman cookery—though apples are often used in combination with cider (see below). Lithuanians season the inside of the goose with caraway seeds and marjoram, then stuff it with chopped sour apples, roasted on a bed of chopped onions, basted with beef or chicken stock. The goose is cut into pieces and served surrounded with large roasted apples.

Cooked apples stand up very well without meat. A Hungarian dish called Apple Cabbage Gul Baba, so named for a leader of the Turkish military force that occupied Hungary, is a sort of cooked coleslaw with apples. A whole, large cabbage is shredded, salted, and left to stand for half a day, then pressed to extract excess water. The cabbage is added to sautéed onions with a spoon of honey or sugar and cooked under cover until not quite tender, at which point a half dozen sweet, chopped apples are added and tossed, and the combination allowed to cook until tender but not soft. To finish, a cup of beef or chicken stock thickened till lightly creamy with flour is added and simmered with the apple-cabbage dish. Salt and lemon juice or cider vinegar can be added to taste.

An Italianish-Californianish risotto can even be crafted from apples. Follow any standard recipe for risotto relying on Italian Arborio rice, then during the last quarter hour of stirring, add one large chopped apple—an Eastern Golden Delicious, Jonathan, or Winesap—*unpeeled*. Finish stirring in remaining stock per standard directions, adding hard, grated cheese at the end with a dash of fresh nutmeg. Usually risotto requires two to two and a half cups of liquid for each cup of rice; instead of the usual combination of white wine and chicken stock, use a good, dry fermented cider.

For the brunch-minded there is apple quiche, this one based on a specialty of Café in the Barn at Seekonk, Massachusetts. Start with a standard butter or shortening pie dough. Combine two-thirds cup chopped leeks, two cups chopped Jonathan, Northern Spy, Winesap, Granny Smith, or Goldrush apples *with peeling*, three-quarters of a pound of soft goat cheese, a cup of heavy cream, three beaten eggs, and a dash of allspice or nutmeg, and bake until the custard is set—about forty to forty-five minutes in a moderately warm oven. A lighter Argentine apple brunch might include *Sopa de Manzanas*, a kind of applesauce soup. As with all applesauces, chutneys, or butters, the object is to blend sweets, tarts, and aromatics. Begin with four medium to large apples of at least three varieties (e.g., Winesap, Jonathan, McIntosh, Cortland, Gala, Grimes Golden, russet, or Northern Spy), core, slice, and *do not peel*; cover with water and stew until tender. Puree, sweeten very lightly if necessary, add a half cup of white raisins, two beaten egg yolks, followed by softly beaten egg whites, and serve either chilled or at room temperature.

Sweet Treats

England, France, and North America have all but defined their national dessert as the apple tart or apple pie. The French prefer an open-faced, glazed tart composed of neatly arranged crescent slices, while the English and Americans bury their apples in sturdy pastry crusts. They are so common they hardly need elaboration here, but for one critical point. Generations of nervous home economics teachers have committed the cruelist blasphemy on the pie-bound apple. In nearly every cookbook now, on almost every newspaper food page, and in gourmet magazines, the first instruction is to "peel, core, and slice" the apple. Unless it's a sauce, coring and slicing is necessary. Peeling, however, robs the apple of both taste and nutrients. Unless the apple you choose has a particularly tough peeling, or you want it to have a perfectly glazed coating on all sides, *leave the peeling alone*! Color and flavor are enormously improved by baking, frying, stewing, or steaming your apples in their skins.

Next to pies and tarts, apple fritters are surely the most universal cooked apple dessert. Almost any apple that holds its shape—York, Golden Delicious, Fuji, Gala, Jonathan, Arkansas Black, Pink Lady, Newtown Pippin—will work well. The object is to cover the apple in some sort of flour-based batter, fry it, then sweeten it. *Pa ssu p'ing kuo*, or caramelized apple, is a classic northern Chinese dessert. Chunky apple slices are rolled in flour, dipped in a mixture of beaten egg white, cornstarch, and a little water, then fried in hot oil until golden brown and drained on paper towels. Once all the apples have been coated and fried, drain the skillet or wok and add two tablespoons sugar and a tablespoon of shortening, then cook until the syrup begins to brown. Using chopsticks or tongs, place the drained fried apple chunks in the bubbling syrup until fully coated, then remove them one by one to an oiled plate. Sprinkle white or black sesame seeds over the caramelized apples, then plunge each chunk into a bowl of ice water to harden and crackle the caramel surface. A slightly lighter version calls for rolling the raw apple chunks in cornstarch first and then again in a water and cornstarch paste directly before plunging them into a bubbling brown syrup made of sesame oil and sugar.

Italian apple fritters come in as many varieties as there are regional cooking styles. Instead of cornstarch, the apple wedges (which in this case should be fully peeled so the batter will adhere all around) are dipped into a standard crepe batter and fried until puffy and golden. A rich holiday version macerates cored apple rounds in a batter made with white wine instead of milk. Instead of being coated with caramelized syrup, the fritters are covered with confectioners' sugar. Dutch Christmas fritters often are made with apple rings that have also been soaked in Calvados or apple brandy. Ligurians are known for a festive winter version they call *Friscieu* in which the apples are chopped and mixed with raisins and currants in a heavier, almost doughlike batter, similar to that used for corn fritters, and then deep-fried.

If fritters are not as common in England or North America, a culinary cousin generally called apple dumplings is. The apples are cored just to the blossom as in baked apples and filled with sugar, butter, a few raisins or currants, and a sprinkling of cinnamon and nutmeg, then wrapped in pie pastry and baked with a quarter inch of water and enough sugar to form a light syrup in the bottom of the roasting pan. In this case, it's wiser to peel the apples to prevent them from popping open and breaking the crust during baking. But save those peelings, and cook them in a simple sugar syrup to make a rose-colored sauce that can be poured over the finished dumplings at the table. The traditional topping for dumplings is hard sauce.

An apple treat common throughout the old Austro-Hungarian territories is called Witches' Froth, or more benignly, Apple Snow. The apples are baked, preferably whole in their skins, then pressed through a food mill or sieve, flavored to taste with sugar (depending on how tart or sweet and how ripe the apples are), vanilla, rum, orange zest, and apple brandy, and gradually folded into stiffly beaten egg whites. If made in advance, the froth should be refrigerated to hold it firm.

Grated or chopped apples can of course be incorporated into cakes or even doughnuts, and cider can be mixed with or substituted for milk. One of the finest and richest apple cakes, common throughout the Appalachians and the American upper South, is Apple Stack Cake, made with dried apples, which bring an exceptionally intense flavor to the cake: Cream together a quarter pound of butter and a half cup of brown sugar, add a half cup of sorghum or molasses and two well-beaten eggs, fold in two-thirds cup of buttermilk; meanwhile mix a teaspoon of baking soda, three and a half cups of flour, a half teaspoon of salt, a teaspoon each of ginger and cinnamon, and a half teaspoon of nutmeg; gradually blend the wet and dry ingredients in alternating quantities and bake in shallow cake tins for half an hour or until done. As the cake is cooking, chop the dried apples and cook for twenty minutes or so in water—or with a little white wine or cider (sweet or hard) or with a dose of rum; sweeten the cooked dried apples *lightly* with cinnamon and/or nutmeg. Once the cakes are cool, spread the dried apple mixture between the layers (it should have the consistency of preserves, just moist enough for a little of the cooking juice to dribble down the side of the cake). Cover the cake and let it sit all afternoon or, better yet, a day.

In Normandy you can find a similar concept, called a *Gateau à la Normande,* in which an apple puree is made of sweetened cider apples and Calvados, then chilled and spread between a dozen or so crepes. Chopped hazelnuts or almonds are sprinkled over the "stack cake," and it is baked for a quarter hour until it's been heated through. The final touch comes at the table when a half cup of hot Calvados is poured over the cake and set afire.

Cider Cooking

Cider was a common base for stews, poached fish, and meats, even for beans and other vegetables, wherever peasants found it better to ferment apples than grapes. There are hundreds of cider dishes common to the cuisines of Asturian Spain, Norman and Breton France, Quebec, and remote parts of New England and the American upper South. The dishes below were selected to illustrate the diversities of cider cookery.

Besugo a la Sidra is a staple of Asturian seafood cooking in which the fish is marinated briefly in garlic, parsley, and lemon juice, then cooked gently with chopped onions, garlic, olive oil, and a glass of cider until it's ready. There are endless variations. Instead of the *besugo*, a kind of porgy, hake is often used, as is snapper or desalinated salt cod, sometimes mixed with shrimp. A flash-cooked salsa of tomato, onion, green pepper, parsley, and even a dash of brandy or white wine can be added to the cider. Scallops and sea urchins are often cooked in cider along the Asturian and Basque coasts.

Rich, garlicky Asturian chorizo sausages are also simmered in local cider with bay leaves to make *Chorizos a la Sidra;* pork cutlets are browned and simmered in cider until they are tender and lightly glazed.

A few hundred miles to the northeast, the French Normans bake sole in a shallow, buttered casserole with a cup of dry cider and then at the very end add several spoonfuls of thick heavy cream. Eel as well is simmered half an hour to forty-five minutes with a half bottle of cider, garlic, shallots, carrots, chopped onion, salt, pepper, and herbs, then removed; the sauce is reduced slightly with a shot or two of Calvados and thickened with a roux before being poured back over the eel on a serving platter.

The Normans are equally famous for braising game in a combination of cider and Calvados. Mapie, the Comtesse de Toulouse-Lautrec, in *La Cuisine de France*, describes spreading a blend of cream and farmer cheese (or other mild creamy cheese) onto the inside of the duck's cavity, seasoning it with salt and pepper, then larding it with pork fat. The duck is browned in butter on all sides and then doused with warm Calvados and flamed. Apple quarters are added to the pan and lightly browned, salt and pepper are added, and a half bottle of cider is poured over the duck before braising it, covered, for an hour or until tender. Two teaspoons of thick cream are added to finish the bird just before serving.

Novelist Annie Proulx, who co-authored perhaps the most complete American guide to hard-cider making, *Sweet and Hard Cider*, gathered many cider recipes, including one she called Cider Squirrels Southern Style, which is similar to double-cooked fried chicken. The squirrels are skinned and cleaned (be sure to remove the smelly scent glands from behind the forelegs, she warns), then soaked in cold salt water, dried, and rolled in seasoned flour, browned in bacon fat, and simmered in dry cider until tender. The pieces of squirrel are allowed to cool, then rolled in the seasoned flour a second time and browned again, this time in butter. The leftover cider, meat juices, and pan scrapings are mixed with a paste made of one tablespoon flour and a cup of cream, which is poured slowly into the pan juices and cooked a few minutes until it becomes a smooth gravy. The same recipe works with rabbit or chicken and is well complemented with either fried or cider-stewed apples.

Proulx also reminds us that the last great stronghold of North American cider making and cider cooking is Quebec, which is largely populated by emigrant Normans. One old standby is baked beans with salt pork, utilizing semidry cider as the baking liquid, but a more distinctive dish, Gaspé-Style Codfish Chowder, comes from Angele Landry-Day's *Le Cidre à boire et à manger*. The chowder starts with a couple-dozen medium mussels, cooked in a cup of cider with a teaspoon of thyme and a coarsely chopped onion. Next, a two-and-a-half-pound cod is skinned and cleaned, and the head, skin, and bones are boiled with a little salt and thyme until the stock is reduced by half; then it is cooled and mixed with the mussel liquid.

To cook the cod, a vegetable broth is first made of two cups chopped celery, two diced onions, a chopped leek, salt, pepper, and bouquet garni plus four cups dry cider, all of which is simmered for half an hour and run through a sieve or food mill and mixed with the mussel and fish stock. Finally, the cod, cut into large chunks, goes into the liquid with a half cup of butter and a half cup of cream and is cooked for fifteen minutes. The mussels are added to the serving bowls with the chowder, and chopped parsley is sprinkled on top.

References

Mary Andere. *Arthurian Links with Herefordshire*. Hereford, U.K.: Logaston, 1996.

Ralph Austin. *Treatise of Fruit Trees* and *The Spiritual Use of an Orchard*. London, 1653. Cited by John Prest. *The Garden of Eden*. New Haven, Conn.: Yale University Press, 1981.

Edward Ayres. *Fruit Culture in Colonial Virginia*. Colonial Williamsburg Monograph, 1973.

L. H. Bailey, Jr. *The Apple-Tree*. New York: Macmillan, 1922.

———. *Field Notes on Apple Culture*. New York: O. Judd, 1886.

S. A. Beach. *The Apples of New York*, vols. 1 and 2. Albany, N.Y.: J. B. Lyon Co., 1905.

John Bultitude. *Apples: A Guide to the Identification of International Varieties*. London: Macmillan, 1983.

Thomas Burford. *Apples: A Catalog of International Varieties.* Monroe, Va.: Burford Brothers, 1991.

Creighton L. Calhoun, Jr. *Old Southern Apples.* Fort Pierce, Fla.: McDonald and Woodward, 1995.

Ricardo A. Caminos. *Late-Egyptian Miscellanies.* London: Oxford University Press, 1954.

R. F. Carlson et al. *North American Apples: Varieties, Rootstocks, Outlook.* East Lansing: Michigan State University, 1970.

P. J. Chapman and S. E. Lienk. *Tortricid Fauna of Apple in New York.* Geneva, N.Y.: New York State Agricultural Experiment Station, Cornell University, 1971.

Paul Correnty. *The Art of Cidermaking.* Boulder, Colo.: Brewers Publications, 1995.

William Coxe. *A View of the Cultivation of Fruit Trees.* England, 1817; repr., Rockton, Ontario, Canada: Pomona Books, 1976.

James Crowden. *In Time of Flood: The Somerset Levels—The River Parrett.* Yeovil, Somerset, U.K.: The Parrett Trail Partnership, 1996.

Hilda Davidson. *Myths and Symbols in Pagan Europe: Early Scandinavian and Celtic Religions.* Syracuse, N.Y.: Syracuse University Press, 1988.

Jean Delumeau. *History of Paradise: The Garden of Eden in Myth and Tradition.* Trans. by Matthew O'Connell. New York: Continuum, 1995.

A. J. Downing. *The Fruits and Fruit Trees of America.* Rev. by Charles Downing. New York: John Wiley, 1870.

Miklos Faust. "The Apple in Paradise." *HortTechnology* (October/December 1994): 338–43.

José Antonio Fidalgo, dir. *Sidra y Manzana de Asturias*. Oviedo, Spain: La Nueva Espana, n.d. (1995?).

D. V. Fischer and W. H. Upsall. *History of Fruit Growing and Handling in the United States of America and Canada*. University Park: Pennsylvania State University, 1976.

Benjamin O. Foster. "Notes on the Symbolism of the Apple in Classical Antiquity." *Harvard Studies in Classical Philology* 10 (1911):39–55.

R. K. French. *The History and Virtues of Cyder*. London: Robert Hale Ltd., 1982.

Robert Graves. *The Greek Myths*. New York: Viking Penguin, 1993.

J. Rendel Harris. "Origin and Meaning of Apple Cults." *Bulletin of the John Rylands Library* 5 (1918–19): 29–74.

U. P. Hedrick. *A History of Horticulture in America*. Repr., Portland, Oreg.: Timber Press, 1988.

Robert Hogg. *The Fruit Manual: A Guide to the Fruits and Fruit Trees of Great Britain*. London: Journal of Horticulture Office, 1884.

H. Frederic Janson. *Pomona's Harvest: An Illustrated Chronicle of Antiquarian Fruit Literature*. Portland, Oreg.: Timber Press, 1996.

James F. Lawrence, ed. *Testimony to the Invisible: Essays on Swedenborg*. West Chester, Pa.: Chrysalis, 1995.

W. A. Luce. *Washington State Fruit Industry . . . A Brief History*. Wenatchee: n.p., n.d.

Eugene S. McCartney. "How the Apple Became the Token of Love." *Transactions and Proceedings of the American Philological Association* 56 (1925).

Kenneth McLeish. *Myth: Myths & Legends of the World Explored*. London: Bloomsbury, 1996.

Walter Minchinton. "Cider and Folklore." *Folklife* 13 (1975): 66–79.

Bruce Mitchell. *Apple City USA: Stories of Early Wenatchee.* Wenatchee, Wash.: The Wenatchee World, 1992.

Joan Morgan and Alison Richards. *The Book of Apples.* London: Ebury Press, 1993.

Vrest Orton. *The American Cider Book.* New York: Farrar, Straus and Giroux, 1995.

Mark Popovsky. *The Vavilov Affair.* Hamden, Conn.: Archon, 1984.

Robert Price. *Johnny Appleseed, Man and Myth.* Gloucester, Mass.: Peter Smith, 1967.

Adele C. Robertson. *The Orchard.* New York: Metropolitan Books, 1995.

Waverley Root. *Food: An Authoritative and Visual History and Dictionary of the Foods of the World.* New York: Simon & Schuster, 1980.

Toby F. Sonneman. *Fruit Fields in My Blood.* Moscow, Idaho: University of Idaho, 1992.

Rick Steigmeyer. *Washington Apple Country.* Portland, Oreg.: Graphic Arts Center Publishing, 1995.

Patrick Sutherland and Adam Nicolson. *Wetland: Life in the Somerset Levels.* London: Michael Joseph, 1986.

Daniel Thomas and Lucy Thomas. *Kentucky Superstitions.* Princeton, N.J.: Princeton University Press, 1920.

Maguelonne Toussaint-Samat. *History of Food.* Trans. by Anthea Bell. Cambridge, Mass.: Blackwell, 1994.

Cuming Walters, ed. *Bygone Somerset.* London: William Andrews, 1897.

Henri Wasserman et al. *La Pomme: Histoire symbolique et cuisine.* Paris: Éditions sang de la terre, 1990.

Henry Weiss. *The History of Applejack or Apple Brandy in New Jersey from Colonial Times to the Present.* Trenton: New Jersey Agricultural Society, 1954.

M. N. Westwood. *Temperate Zone Pomology,* 3rd ed. Portland, Oreg.: Timber Press, 1993.

L. P. Wilkinson. *Bulmers: A Century of Cider-Making.* Hereford, U.K.: David & Charles, 1987.

R. R. Williams. *Cider and Juice Apples.* Bristol, U.K.: Long Ashton Research Center, 1988.